WEATHER
A FORCE OF NATURE

Spectacular images from
Weather Photographer of the Year

CSIRO PUBLISHING

RMetS
Royal Meteorological Society

First published by the Natural History Museum,
Cromwell Road, London SW7 5BD

A catalogue record for this book is available from the National Library of Australia.

ISBN 9781486315727

This edition published exclusively in print only, in Australia and New Zealand, by CSIRO Publishing.

CSIRO Publishing, Locked Bag 10
Clayton South VIC 3169, Australia
Telephone: +61 3 9545 8400
Email: publishing.sales@csiro.au
Website: www.publish.csiro.au
Sign up to our email alerts: publish.csiro.au/earlyalert

10 9 8 7 6 5 4 3 2 1

Editor Hannah Mallinson
Caption writer Matthew Clark
Image grading Stephen Johnson www.copyrightimage.com
Colour proofing Saxon Digital Services UK
Printing Toppan Leefung Ltd, China

p.5: Supercell twister, near Wray in northeastern Colorado, USA by Dusty Dhillon (see p.23)

pp.8–9: Sparkling jewels of ice, Lake Baikal, Russia by Alexey Trofimov (see p.96)

Acknowledgement
CSIRO acknowledges the Traditional Owners of the lands that we live and work on across Australia and pays its respect to Elders past and present. CSIRO recognises that Aboriginal and Torres Strait Islander peoples have made and will continue to make extraordinary contributions to all aspects of Australian life including culture, economy and science. CSIRO is committed to reconciliation and demonstrating respect for Indigenous knowledge and science. The use of Western science in this publication should not be interpreted as diminishing the knowledge of plants, animals and environment from Indigenous ecological knowledge systems.

CONTENTS

Foreword

DAVID GRIGGS

The photographs in this book are a celebration of the Weather Photographer of the Year competition, which has been run annually by the Royal Meteorological Society since 2016, with the aim of finding the very best photographs from around the world that depict weather in its widest sense. A highly-respected panel of judges, including meteorologists, photographers and photo editors look for images that combine photographic skill with meteorological observation. Winners are chosen by the judging panel and shortlisted entries are also put to a vote to discover the public's favourite.

As well as the main competition, there is also the Young Weather Photographer of the Year category (for photographers 17 and under only) to encourage and recognize the younger generation of weather photographers.

Free to enter, the competition has become increasingly popular, with the number of entries growing every year from just 800 in 2016 to more than 7,500 in 2020. Photographs are submitted from all around the world and photographs from 29 countries are featured in this book.

The book includes a selection of photos from the first five years of the competition. They illustrate perfectly the beauty to be found in the phenomena that bring us our weather, that supports all life on Earth. It is hard to pick favourites because every photograph is both beautiful to look at and interesting meteorologically. But the editors have twisted my arm and so, because I love walking in the countryside on a cold crisp winter's day, my first pick would be *Frost and Ice* by Julie Fryer. I was lucky to live in Australia for 10 years and work with aboriginal people and to visit Uluru, so my second pick is *Strikes on Uluru* by Christoph Schaarschmidt. It is such a special place and I also love lightning.

But, our weather is changing. Climate change is now widely recognized as one of, if not the most serious threat to our future. Climate is weather averaged over a period of time, usually 30 years, so, when we talk about climate change what we are really talking about is changing weather because, as human beings (or any other living being), we do not experience climate, we experience weather. So, if the climate changes this means the weather we experience changes. In other words, climate is what you expect but weather is what you get.

So, to highlight our changing weather and the threat posed by climate change the photographs in this book are accompanied by six essays on various aspects of climate change. When thinking about climate change the first questions that we need to try and answer are, is the climate changing and if the climate is changing what is causing it. These questions are covered in the first essay, and are followed by an essay looking at how the changing climate will affect our weather and, in particular, extreme weather events, such as heatwaves and storms. There are then two further essays covering specific regions of our planet, namely the oceans and the cryosphere (snow and ice), and then a fifth which looks at how climate change is transforming our natural world. Last, but not least, the final essay discusses global responses to climate change – what we are already doing and the additional actions we can take to tackle it.

I hope that you enjoy these stunning photographs, but that you also take the time to read the essays, because we all have a responsibility to understand the risk climate change poses to the world and we all have a part to play in limiting it and its impacts. In that way, future generations will also be able to enjoy and photograph these beautiful weather phenomena for themselves.

How and why Earth's climate is changing

RICHARD ALLAN

We can monitor our changing climate through local measurements taken at ground-based meteorological stations, by weather balloons or aircraft sampling the atmosphere, or by ship measurements and automated buoys that sample the ocean to depth. Remote sensing by interpreting the electromagnetic fingerprints measured by satellites further provides continuous, global coverage of the planet. We also know how the Earth's climate has changed in the distant past based on indirect or 'proxy' data such as tree rings, ice cores and fossil or other ancient geochemical evidence. This tells us that the Earth's climate has always changed: throughout the billions of years of the planet's history as the atmosphere and life co-evolved, across hundreds of millions of years as continents crept across the surface, and over tens of thousands of years as predictable and periodic oscillations and wobbles in the Earth's orbit around the sun set in motion glacial cycles. These changes over vast periods of time are not readily perceptible to us and yet provide insight and context for current changes, which we know are caused by (and are increasingly damaging) human societies.

A wealth of knowledge can be gleaned from kilometre-deep cores into the ice of Antarctica, like million-year time machines that capture colossal fluctuations in Earth's climate. Fingerprints from the composition of the ice and tiny bubbles trapped within it, can tell us about swings in temperature from the last mild interglacial period about 125,000 years ago, followed by a long glacial period that was finally punctuated by our current interglacial climate about 12,000 years before present. Combining with other geochemical evidence we know that during the last glacial maximum, about 21,000 years ago, the planet's surface temperature was 5–6°C colder than in pre-industrial times. Vast amounts of ocean waters were locked away in giant ice sheets over North America and Eurasia leading to sea levels around 120 m lower than today.

It has long been known that cyclical changes in Earth's orbit can explain the timings of the swings between glacial and inter-glacial climates over the past few million years. These changes alter the distribution of sunlight across the surface of the planet and how much this heating contrasts between winter and summer. The orbital changes set in motion a chain of events causing ice sheets to expand and contract. But the fossilized records of the atmosphere, trapped in tiny bubbles in the ice cores, show how the concentrations of the greenhouse gases, carbon dioxide and methane also rise and fall with temperatures in conjunction with glacial to interglacial swings. And only with the helping hand of the changing greenhouse effect and coverage of reflective ice can the magnitude of glacial climate cycles be explained.

The ice core records show how carbon dioxide has varied over the past 800,000 years, from less than 0.02% of the atmosphere (below 200 ppm) during frigid periods up to about 280 ppm during mild interglacial interludes. Yet current concentrations are 415 ppm, well outside the natural fluctuations. Until now these levels have never been experienced by anatomically modern humans, who first appeared over 200,000 years ago. It is well known that carbon dioxide emissions (mostly from fossil fuel burning but also by changes to land cover including deforestation) are causing levels to rise in the air we breathe. Although these emissions are small compared to the natural flow of carbon dioxide into and back out of the atmosphere each year, as the planet 'breathes' through growth and decay of vegetation over an annual cycle, they tip the balance such that concentrations inexorably rise by 2–3 ppm each year.

The Earth's greenhouse effect naturally insulates the planet, keeping it habitable by reducing how effectively the planet loses the energy gained from the sun by emitting infrared heat back out to space. But this warming influence is being amplified as greenhouse gas concentrations rise. Research stemming from the 19th century determined how greenhouse gases absorb and emit infrared radiative heat; this cornerstone of climate science has developed over the years into a fundamental understanding of how the climate system works, and today satellite measurements confirm this emerging greenhouse gas fingerprint. Rather than naturally responding to and amplifying climate change, as is the case over many thousands of years across glacial cycles, today, human-caused greenhouse gas emissions are driving Earth's climate into an uncharted, much warmer future.

The average temperature of the planet has already risen more than 1°C since industrialization was well underway in the 19th century. The planet has not warmed steadily since the early 20th century – there are fluctuations as the sun varies in its intensity; or following major volcanic eruptions that temporarily block out sunlight; or as the oceans fluctuate naturally from one decade to the next. These factors explain a barely perceptible and temporary slowing of global warming at the beginning of this century. A longer pause in warming in the 1950s and 1960s is partly explained by increased particle pollution that scattered more sunlight back to space and pulled against the warming greenhouse gas influence. Yet, since the 1970s, each decade has been significantly warmer than the one before it, as the heating influence of the growing greenhouse effect has dominated. Additional signals of climate change come from declining Arctic ice, melting glaciers, rising sea levels and increasing ocean heat. Precipitation is changing, as wind patterns shift with the chaotic fluctuations in the weather; but these weather patterns are increasingly being nudged out of kilter by the human-induced warming and the smaller but more uneven cooling effects of aerosol particle pollution. Atmospheric moisture is increasing too, amplifying the warming through its strong greenhouse effect but also intensifying heavy rainfall.

To predict how climate change will unfold requires the understanding of past changes. The most up-to-date knowledge surrounding the physics of our atmosphere, ocean and land is applied across millions of grid boxes that make up a three-dimensional jigsaw model planet, represented by millions of lines of computer code. These global climate simulations are virtual laboratories with which to test hypotheses and answer 'what if' questions such as: how much of the observed warming can be explained by greenhouse gas increases alone; would the flooding event have occurred if ocean temperature had been at pre-industrial levels?

Different climate simulation models have been developed over many decades at multiple science centres across the world. These are our primary tools for making projections of the future, applying plausible scenarios of socioeconomic development leading to continued high greenhouse emissions or the drastic cuts in greenhouse gas emissions needed to stabilize climate. These future projections show that without rapid, sustained and widespread cuts in emissions across all sectors of society, warming of the planet will continue beyond levels considered dangerous. Carbon dioxide levels are already about 50% higher than pre-industrial levels; at levels 100% higher than (or double) the pre-industrial levels a global-mean warming of about 3°C is a reasonable expectation in the long term based upon the latest science. This is a staggering change in climate, considering it is about half as big as the global temperature change that took Earth from glacial conditions 21,000 years ago, when ice sheets covered northern landmasses, to today's mild climate.

The science shows that warming is greater over land areas and in particular over the Arctic where melting of bright, reflective sea ice uncovers dark oceans, causing more sunlight to heat the surface and amplify warming. Intense and sustained rainfall events with associated flooding are already becoming more severe as the air hangs heavier with moisture that fuels the storms. Conversely, when weather patterns generate heatwaves and drought these also become more severe as the temperatures climb further and the atmospheric thirst for water increases. Sea levels will continue to rise far into the future as the oceans steadily take up heat and expand, and land ice melts, adding more water to the sea.

Our extensive scientific knowledge about how and why climate is changing empowers humans to make informed decisions that will limit ongoing climate change and its effects on societies. There is therefore a window of opportunity to limit climate change by drastically cutting greenhouse gas emissions and to plan for adaptations that reduce the impact of future climate change that cannot be avoided.

Our climate and extreme weather

MIKA RANTANEN

Although climate change refers to relatively slow, long-term changes in the climate, it can have significant impacts on our everyday lives. Regionally, we can see changes such as European summers getting hotter or Arctic winters becoming milder. By definition, if the climate is changing then the weather must be changing too, and one implication of this change is that weather phenomena that were previously considered extreme are becoming more common.

The average temperatures at the Earth's surface have risen by just over 1.0°C (1.8°F) since pre-industrial times, largely due to increased levels of greenhouse gases in the atmosphere as a result of human activity. This is the average across the whole Earth, including both ocean and land. The continental land areas – where most people live – have warmed already more than 1.5°C (2.7°F). Global trends in other meteorological averages such as rainfall, sunshine and wind are less obvious. However, one thing is becoming clear – many weather extremes are changing in frequency or intensity across the globe, and this is increasingly being attributed to human-induced climate change. Through using historical weather observations and state-of-the-art models, scientists are able to simulate a world with and without anthropogenic climate change to calculate how much more likely an event has become. This rapidly developing area of research is called climate change attribution.

Out of all extreme weather phenomena, changes in temperature-related extreme weather, particularly high temperature events such as heatwaves, can be connected to climate change most confidently. Heatwaves can occur when a strong high pressure area in the upper troposphere stalls and remains in place for several days, up to weeks. The increased global average temperatures due to climate change mean that heatwaves have become more intense, frequent and long-lasting. Examples of recent exceptional heatwaves in which human-induced climate change has been found to have a clear contribution are the 2019 European heatwave and the 2020 Siberian heatwave. During these events, France recorded its all-time highest temperature of 46.0°C (114.8°F) in June 2019 and Verkhoyansk, a town north of the Arctic Circle in Siberia, reported 38.0°C (100.4°F) in June 2020. Both events would have been highly unlikely to take place with similar magnitude without human influence.

In many parts of the world, climate change is making droughts worse. Higher temperatures enhance evaporation and so regions which are naturally dry can become even drier. This can lead to a positive feedback loop where, as the soils become drier, the ground is cooled less by evaporation, raising surface temperatures even further. Record-high temperatures combined with drought conditions have been the cause of recent extensive wildfires in areas such as the western USA, Siberia and Australia. While the proportion of naturally started versus human-started wildfires varies greatly by region, research suggests that globally around 90% of wildfires are triggered by humans. However, one thing is clear: climate change is making wildfires worse.

Such wildfires have various effects on weather and air quality. Out of the 3.3 million premature deaths that occur each year because of poor air quality, research suggests wildfires are responsible for around 5% of them. Wildfires can also trigger pyrocumulus clouds, which form when the hot air above the fires rises up, cools and forms cloud droplets. Lightning strikes from these clouds can then ignite further fires. In some cases, because they have very intense updrafts, pyrocumulus clouds reach the stratosphere by punching through the tropopause layer and in doing so inject a large quantity of smoke particles to levels where they can linger for months, similar to moderately explosive volcanic eruptions.

In terms of cold spells, we are still experiencing them but they are generally occurring less often and are weaker in intensity. We are seeing a decrease in the number of low temperature records being broken, while there is an increase in the number of high temperature records being broken. One area that is warming three times faster than the global average is the Arctic, which we call Arctic amplification. One of the main drivers for this warming is the loss of sea-ice. When the sea-ice melts during the summer months, areas of darker open ocean are exposed, which absorb more heat from the sun. This leads to further ice melting and a self-reinforcing cycle begins. Due to the declining temperature difference between the poles and lower latitudes, climate change could weaken Earth's jet streams, which would have consequences for weather systems at the surface. The links between climate change, jet streams and weather patterns are still very uncertain and are therefore the subject of active research to understand them better.

Water evaporates more easily in warmer air. So, as the climate warms, the atmosphere will on average contain more water vapour, meaning it can produce more intense precipitation which may lead to flooding. For example, Hurricane Harvey produced over 1,500 mm (60 in) of rainfall over southern Texas in August 2017, which was the highest tropical cyclone rainfall amount ever recorded in United States history. An attribution study has shown that the total rainfall was three times more likely and 15% more intense due to human-induced climate change.

Tropical cyclones are formed when warm, moist air over the ocean rises up and induces an area of lower air pressure below. The surrounding air then flows into the area of lower pressure, creating a swirling cyclone where warm and moist air rises up in the centre of the system. The energy source of tropical cyclones is the warm sea surface, which needs to be at least 26.5°C (79.7°F) to a depth of at least 50 m (164 ft). Rising sea surface temperatures thus provide more fuel for tropical cyclones. Indeed, it has been discovered that the speed with which these systems strengthen has increased in recent decades. The rapid intensification of tropical cyclones can pose a major problem for coastal cities because weather forecast models still struggle to predict the intensification rate accurately. An underestimation of a rapid intensification can hinder preparations in populated areas and thus risk additional damage to society. Besides intensifying faster, the probability of tropical cyclones reaching higher categories has also increased in recent decades; this is consistent with the warming of sea surface temperatures.

Climate change is expected to make the impacts of coastal tropical cyclones worse due to rising sea levels. The global average sea level has risen by about 18 cm (7 in) since the beginning of the 20th century, driven by the thermal expansion of warming oceans and the melting of glaciers and ice sheets. The coastal flooding associated with tropical cyclones is caused by both heavy precipitation and a storm surge – a weather-generated sea level rise due to strong winds and low pressure. Thus, the flooding by tropical cyclones is being enhanced by both increased moisture content in the atmosphere and the long-term rise of the sea level due to global warming.

Mid-latitude cyclones, which form along frontal boundaries as a result of diverging air aloft and converging air at the surface, are also becoming wetter. One recent example is December 2015 when the UK set a new 24-hour rainfall record during Storm Desmond. The rainfall was associated with a phenomenon called an atmospheric river, a narrow corridor of atmospheric moisture in the lower troposphere, moving from tropical or subtropical areas towards the mid-latitudes. The evidence suggests that increased atmospheric moisture content will enhance the intensity of precipitation related with atmospheric rivers, and as a consequence hydrological extremes.

In summary, there is growing evidence that many extreme weather events are becoming more frequent or intense in a warming climate. The most confident attribution research applies to the increasing intensity of heatwaves, but worsening droughts, more extensive wildfires, more intense rainfall and faster intensification of tropical cyclones also belong in the long list of extreme weather impacted by climate change.

Wall of dust

Tina Wright

The leading edge of a haboob races across the parched landscape of Arizona, USA. Haboobs are a special type of dust storm associated with thunderstorms. As the downdraft from the thunderstorm hits the ground it spreads out, resulting in strong and gusty outflow winds. This effect is particularly marked in thunderstorms containing a microburst or downburst – a column of intensely sinking air that can reach great speeds as it nears the ground. In arid regions the resulting outflow winds lift dust particles, lofting them to heights of 1,500 m (5,000 ft) or more. The leading edge of the dust-filled air becomes well-defined due to converging winds near the storm's gust front (the leading edge of the outflowing cold air), resulting in a spectacular wall of dust. As the leading edge of the storm advances it is capable of reducing visibility to a few metres in a matter of seconds.

Nikon D850 + Tokina AT-X 16–28 mm *f*/2.8 PRO FX: 1/15 s; *f*/4.5; 16 mm; ISO-400.

Lighting up the night

Hugo Begg

A single cloud-to-ground lightning bolt arcs to Earth, passing through dense cloud layers in a thunderstorm at Trial Bay, Australia. 'After watching this spectacular lightning show for an hour I was running out of time to achieve the desired shot,' says Hugo. 'Not two minutes after this photograph was taken, the full force of the storm hit with heavy rain and gale force winds, making any further photography impossible.' Two types of lightning are visible in this shot: cloud-to-ground lightning, and so-called 'sheet' lightning (the flash of brightness, top left), which occurs when the lightning channel is embedded well within the storm cloud, or is masked by an intervening layer of cloud, such that the light is scattered in all directions to produce a diffuse flash. Sheet lightning can also occur when the lightning is many kilometres away over the horizon.

Pentax K-70: 30 s; *f*/4.5; 40 mm; ISO-100.

COMMERCIO

Rolling shelf before the storm

Sandro Puncet

A stunning arcus cloud rolls across the Croatian coastline near Mali Losinj. Arcus clouds form at the gust front of thunderstorms, which is the leading edge of the rain-cooled air flowing out from the storm. The best examples usually occur when individual storm cells merge into larger clusters and lines of thunderstorms. The rain-cooled air from individual storms then combines into a much larger region of cold, dense air that can help to propel the area of storms forward as it spreads outwards. Arcus clouds that are as well-formed and extensive as the one photographed here are often accompanied by damaging wind gusts.

Desert planet

Dennis Oswald

Strong winds whip up a sandstorm in the Mesquite Dunes of Death Valley National Park, California, USA. Sandstorms occur when strong winds pick up loose particles of sand and dust when the land surface is very dry, especially where vegetation is sparse. As the wind increases in strength, individual grains of sand first begin to vibrate. With further increases in the wind speed, they begin to bounce along the surface in a process known as saltation. As they impact the surface, small particles break off which may then be lofted to greater heights as they become suspended in the air. Assuming a plentiful supply of sand, the movement of the sand grains eventually and collectively results in the development of sand dunes, which slowly migrate across the landscape in the direction of the prevailing wind.

Nikon D800E + 70–200 mm *f*/2.8: 1/80 s; *f*/9; 125 mm; ISO-100.

Three-pronged strike

Steven Genesin

Multiple cloud-to-ground lightning bolts illuminate shafts of precipitation falling from a high-based evening thunderstorm offshore of Brighton Jetty, Adelaide, Australia ('high-based' meaning that the main base of the thundercloud is located higher than about 1,500–2,500 m (4,900–8,200 ft) above ground level). Steven explains, 'A summer storm approached the coast and, as the sun set, lightning strikes started to occur in the distance behind the jetty. Although I went to the beach to take shots of the sunset, in the end the storm created a much better subject to focus on!'

Olympus E-M10 + Olympus M.17 mm *f*/1.8: 1/2 s; *f*/9; 17 mm; ISO-200.

Mountain thunderstorm

Robert Juvet

Pink-tinged lightning illuminates the mountains near Appenzellerland, Switzerland, during a powerful nocturnal thunderstorm. Robert says, 'This photograph was taken in front of my house in Appenzellerland, Switzerland. As the storm developed I stood on my terrace to watch and photograph the spectacle. The flash frequency was quite high, which meant that I could work with relatively short exposure times. I particularly like this shot because of the ambient light in the clouds, and the beautiful pink coloration of the lightning which provides contrast with the green of the landscape'.

Nikon D810 + Tamron SP 15-30 mm f/2.8 Di VC USD A012N: 8 s; f/22; 28 mm; ISO-100.

Supercell twister

Dusty Dhillon

A tornado descends from the rotating base of a large supercell thunderstorm near Wray in northeastern Colorado, USA. Because it is generally not possible to obtain a direct measurement of the extreme wind speeds within a tornado, the intensity of tornadoes is classified using a system called the Enhanced Fujita (EF) Scale, which is based entirely on the damage caused by the tornado. Damage of a given severity may then be related to an estimated range of likely wind speeds. The EF-Scale runs from 0 (least intense) to 5 (most intense). Tornadoes of intensity EF1, for example, produce wind speeds in the estimated range of 138–177 kph (86–110 mph), whereas EF4 tornadoes produce winds in the estimated range 267–322 kph (166–200 mph). The Wray tornado of 8 May 2016, photographed here, was rated as a high-end EF2, with estimated wind speeds as high as 217 kph (135 mph). In the UK and some parts of Europe, a different intensity scale – the International Tornado Intensity (T) Scale – is used which runs from T0 (least intense) to T10 (most intense).

Nikon D800 + 28–300 mm *f*/3.5–5.6: 1/320 s; *f*/8; 58 mm; ISO-1000.

Evening mammatus

Boris Jordan

A spectacular array of mammatus clouds glow red and purple in the evening sunlight near Leipzig in Germany. Mammatus clouds comprise a series of pouches hanging from the base of a cloud, resembling the udders of a cow. They typically develop on the underside of the anvil of a mature thunderstorm. Although the mechanism of formation is still a topic of some debate, it is thought to involve sinking pockets of colder, denser air within the cloud which eventually protrude from the rest of the cloud mass to form the characteristic mammatus lobes. Mammatus clouds also occasionally form on the underside of other types of cloud such as altocumulus. They are especially visible when low-angle sunlight around sunrise or sunset catches the individual cloud lobes so that they stand out in relief.

Canon EOS 6D + Canon EF 16–35 mm *f*/4L IS USM: 1.3 s; *f*/5.6; 16 mm; ISO-640.

Steering clear of trouble

Marc Brown

A mature cumulonimbus cloud as seen from the air above the Golfe du Lion near Montpellier, France. Cumulonimbus clouds are one of the primary meteorological hazards to aviation. Strong currents of rising and falling air in the cloud may be associated with severe turbulence, whilst large hail and severe icing can also be encountered in the more intense storms. For this reason, commercial aircraft often reroute in order to avoid passing through these clouds. The typical height of the top of a mature cumulonimbus over land in the summer season is around 12–14 km (7½–8½ miles) above ground level, which compares to a typical cruising altitude for larger commercial aircraft of around 11 km (6¾ miles).

Nikon D5600: 1/500 s; *f*/11; 34 mm; ISO-200.

Hail curtain

Mark Boardman

A winter hailstorm approaches the Jodrell Bank Observatory near Macclesfield in northwest England. When air masses of polar origin move equatorward over relatively warm waters they become unstable, resulting in the development of showers and thunderstorms. These showers frequently affect coastal locations exposed to northwesterly or northerly winds (southwesterly to southerly winds in the southern hemisphere) and, although generally less powerful than continental convection in the summer season, they are capable of producing severe weather. When the height in the atmosphere at which temperatures fall below freezing (known as the freezing level) is low enough, these showers fall as a mix of snow and soft hail (also known as graupel). Frozen precipitation is relatively large and optically opaque, and so these wintry showers often produce spectacular curtains of precipitation that sweep across the landscape as they move onshore.

Nikon D3300: 1/320 s; *f*/11; 55mm; ISO-400.

Painting in the sky

Simon Anderson

A complex array of lightning channels streak across the sky above Eastbourne Pier on the south coast of England during an overnight thunderstorm. Thunder occurs as a result of intense heating of air surrounding the lightning channel. The heated air expands rapidly, which sets up shock waves that radiate outwards in the form of sound waves that we hear as thunder. When a bolt of lightning strikes close by, the thunder is sudden and may sound almost like an explosion, but with increasing distance the sound waves tend to disperse, leading to the rumbling sound that is perhaps more familiar. The sound waves from different parts of the lightning channel, at greater or lesser distances from the observer, take varying amounts of time to reach the observer, which also explains why thunder tends to last several seconds and why a single roll of thunder often comprises multiple peaks and troughs in volume. Thunder is typically audible up to about 16–24 km (10–15 miles) from the lightning, though this varies according to the atmospheric conditions.

Nikon D750 + Tamron SP 15–30mm *f*/2.8 Di VC USD A012N: 30 s; *f*/7.1; 18 mm; ISO-250.

A pink dusting

Tina Wright

On 9 July 2018 a line of severe thunderstorms moved from east to west through the state of Arizona, USA, producing one of the largest haboobs (dust storms) on record. It measured almost 1½ km (1 mile) high and travelled nearly 320 km (199 miles), crossing the border into California after dark before dying out. This panoramic image was taken near Yuma, Arizona, where the leading edge of the dust storm formed a wall that stretched almost from horizon to horizon as it approached at sunset. The last rays of the evening sun give the haboob its beautiful pink hue. The close association of haboobs with thunderstorms is self-evident in this example, given the presence of a spectacular shelf cloud – the dense roll of cloud that forms along the leading edge of a thunderstorm – which can be seen to fill much of the sky above the advancing wall of dust.

Nikon D850 + Tokina AT-X 16–28 mm *f*/2.8 PRO FX: 1/20 s; *f*/5; 16 mm; ISO-100.

Whirling devil

Hadi Dehghanpour

This photograph captures the moment a powerful dust devil hits a group of tents in a village near Noush Abad in Iran. Dust devils are relatively small, short-lived whirlwinds occurring during sunny conditions when the air in contact with the ground is heated strongly. This heating creates shallow convection currents that tend to organize into small regions, or cells, of ascending and descending air near ground level. Areas of weak rotation, associated with local variations in the wind speed and direction, can be strongly focussed and amplified as air converges in towards the areas of ascending air. This process occasionally results in the development of a well-defined column of strongly rotating air, and a dust devil is born. Although most dust devils are harmless, they occasionally grow large and strong enough to constitute a hazard to outdoor activities, as in this example. Dust devils resemble tornadoes, in so much as they are a strongly rotating column of air. However, they are almost always weaker than tornadoes and tend to be much shallower. Crucially, whilst the rotation in dust devils generally only extends a few tens of metres to perhaps 100–200 m (328–656 ft) above ground level, tornadoes are connected to the base of a convective cloud aloft.

Canon 5D Mark III + Canon 75–300 mm: 1/400 s; *f*/11; 120 mm; ISO-100.

Beating in vain

Tohid Mahdizadeh

Bush and forest fires, such as this heathland fire in northwest Iran, can be a major hazard in arid and semi-arid regions of the world. Fires may start as a result of human activity, or due to natural processes such as lightning strikes, the latter being especially likely where storms are high-based and relatively little precipitation reaches the ground. The likelihood and severity of wildfires is strongly influenced by the weather conditions. In particular, high surface temperatures and low humidities dry out vegetation and increase the likelihood of ignition, whilst strong winds act to fan the flames, aiding the spread of existing fires.

Waterspout party

Sandro Puncet

A rare spectacle unfolds as multiple waterspouts develop underneath the base of a line of storms over the Adriatic Sea, as seen from near Mali Losinj, Croatia. Waterspouts are simply tornadoes occurring over water. They are particularly common in the autumn and early winter months over the Mediterranean Sea and adjoining water bodies, when outbreaks of cold air spread over the still-warm waters, creating strong instability. In this example, the fact that the waterspouts are arranged in a line suggests that they have developed along a well-defined convergence zone – a boundary where winds of differing speeds and directions meet. As the air converges it is forced to rise, creating a line of towering clouds. The differing wind speeds and directions generate spin along the boundary which, when ingested by the rising air and stretched in the vertical, may be amplified to tornadic strengths, resulting in development of the waterspouts.

Strikes on Uluru

Christoph Schaarschmidt

Multiple forks of lightning strike as torrents of water cascade down from a rain-soaked Uluru in Australia, during a high-based desert thunderstorm ('high-based' meaning that the main base of the thundercloud is located higher than about 1,500–2,500 m (4,900–9,200 ft) above ground level). Christoph says, 'I'd heard a lot about how beautiful Uluru looks in the rain. This is one of the reasons why Uluru is such a special place for the Anangu – the local Aboriginal clan. I never believed I would see it with my own eyes, given how rare rainfall is in the arid, red centre of Australia'.

Canon EOS 70D + 11–16 mm: 5 s; *f*/13; 16 mm; ISO-100.

Evening strike

Julian Elliott

An evening thunderstorm rolls across the lavender fields on the Plateau de Valensole in Provence, France. Thunderstorms require three ingredients to form. Firstly, instability. This occurs when the temperature decreases rapidly with height in the atmosphere, such as when there is strong heating of the surface on a hot summer's afternoon, or when cold air aloft overspreads warm air near the surface. Secondly, moisture. Moisture fuels the storm because as air rises in the storm's updraft, condensation of water vapour releases heat which helps to drive the updrafts to greater heights. Thirdly, a trigger. Storms require a lifting mechanism to trigger the updraft – for example, where the air is forced to rise over hills or mountains, or along boundaries such as sea breeze fronts.

Canon EOS 6D + EF 28–70 mm *f*/2.8L USM: 3/5 s; *f*/8; 50 mm; ISO-1600.

Mammatus highway

Dennis Oswald

Mammatus clouds on the edge of a severe thunderstorm in the Texas Panhandle, USA, catch the very last rays of evening sunlight. A common misconception is that the presence of mammatus signifies that a tornado is about to form. Although mammatus clouds are usually associated with well-developed cumulonimbus clouds, which can produce tornadoes on occasion, the vast majority of cumulonimbus clouds do not produce tornadoes, which means the presence of mammatus is not a reliable indicator of tornadic activity in itself. The mammatus may, however, be a more reliable indicator of the presence of other hazards associated with thunderstorms such as severe turbulence and lightning.

Nikon D800E + 16–35 mm *f*/4.0: 5/2 s; *f*/5.6; 16 mm; ISO-320.

Storm on a shelf

Maja Kraljik

A shelf cloud surges across the coastline at Umag in Croatia. A shelf cloud, or arcus, is a low-level, horizontal, wedge-shaped cloud that occurs along the leading edge of the rain-cooled air flowing out of a mature thunderstorm. Warm, moist air is forced to rise abruptly above the rain-cooled air, causing the moisture to condense into one or more dense rolls of cloud. As the shelf cloud passes overhead, it is accompanied by gusty winds and a sharp fall in temperature, heralding the arrival of the storm.

Nikon D40: 1/125 s; *f*/8; 18 mm; ISO-200.

Devils Tower supercell

John Finney

A supercell thunderstorm approaches Devils Tower in Wyoming, USA. The blue-green coloration in storm clouds is often attributed to hail. However, a more likely explanation concerns the scattering of light by both water droplets and solid ice (including graupel and hail) within the storm cloud. In deep storm clouds, which have a large water mass, multiple scattering yields a large integrated path of light through the cloud water and ice. Since blue light is transmitted in water and ice, whilst red light is absorbed preferentially, white light becomes blue provided that it passes through a sufficiently long path of water or ice (this effect also explains the blue coloration of large blocks of ice, as discussed on p96). When this blue light is combined with reddened sunlight, for example close to sunset or sunrise, the resulting colour may be perceptually green. An alternative theory is that the storm cloud provides a dark backdrop against which the light scattered by molecules and tiny suspended particles in clear air may be seen. This scattering yields blue or green 'airlight'. However, the coloration is only evident when the airlight is seen against a very dark backdrop, such as a dense storm cloud.

Nikon D810 + 14–24 mm *f*/2.8: 1/15 s; *f*/11; 19 mm; ISO-64.

Classic strike

Dennis Oswald

A powerful cloud-to-ground lightning bolt strikes near the interface between the rotating updraft and the downdraft region of a supercell thunderstorm in far southwestern Oklahoma, USA. Over the Great Plains of the central USA, supercell thunderstorms frequently develop during a well-defined storm season that runs approximately from April to June, peaking in May. On 3 May 1999, a powerful supercell thunderstorm spawned a violent tornado that caused devastation as it tore through parts of central Oklahoma. A mobile weather radar measured a wind speed of approximately 484 kph (301 mph) as the tornado ripped through the community of Bridge Creek in Grady County. This remains the strongest wind speed ever recorded in a tornado.

NIKON D800E + 16–35 mm *f*/4.0: 1/6 s; *f*/9; 24 mm; ISO-50.

Lighting up the sea

Elena Salvai

An intense cloud-to-ground lightning bolt strikes the sea surface below the cliffs at Riomaggiore, Italy. Lightning is a large electrical spark produced by electrons moving from one place to another. The electrons are moving so fast that the air around them glows – this is what we see as the lightning flash. The lightning flash ends when the regions of positive and negative electric charge within the thundercloud, or between the cloud and the ground, are neutralized, so that no more electrons flow. Lightning normally takes the shortest and quickest route to the ground, commonly striking tall objects such as trees and skyscrapers. However, this is not always the case. Lightning can also strike open ground or, as in this case, open water.

Canon EOS 5D Mark IV + EF 24–70 mm *f*/2.8L II USM: 46 s; *f*/5; 24 mm; ISO-100.

Thirsty earth

Abdul Momin

There are four main types of drought: agricultural, hydrological, meteorological and socio-economic. Although there is no universal definition for meteorological drought, it is often defined as a given period of time (months or years) where rainfall is abnormally lacking. Currently, 80% of the annual rainfall in Bangladesh occurs during the monsoon months (June–September), with the rest falling in the remaining eight months. As a result, water levels drop and cultivable lands become dry during winter, meaning in some areas farmers cannot use water pumps to irrigate and people use fields as shortcuts from one village to another. Although the dry season is not technically a drought, because the lack of rain is a regular and relatively predictable occurrence, climate models indicate that the rainfall variability could shift under climate change, with more intense rainfall during the monsoon months and even less in the winter months, so droughts (abnormal lack of rainfall) become more common during the dry season.

Hasselblad L1D-20c: 1/400 s; *f*/3.5; 10.26 mm; ISO-100.

After the fire

Dmitriy Kochergin

An unearthly scene is encountered in the aftermath of forest fires in the Bizhbulyaksky District of Bashkortostan, Russia. Although it is often difficult to attribute individual weather events to climate change, current projections show that it is very likely that temperature and rainfall extremes will increase in a warming world. Where high temperatures coincide with unusually prolonged or acute droughts, forest fires are all too often the catastrophic result. Wildfires also contribute to global warming in themselves, because they release large amounts of carbon dioxide – a greenhouse gas – into the atmosphere, and because they destroy vegetation that would otherwise have helped to remove carbon dioxide by photosynthesis.

Fujifilm X-T2 + XC 50-230 mm *f*/4.5-6.7 OIS: 1/2200 s; *f*/5.6; 63.2 mm; ISO-400.

Red terror

Tori Jane Ostberg

A large tornado churns up a huge, rotating cloud of red dust and debris near Wray in northeastern Colorado, USA. A tornado is a violently rotating column of air that extends between the base of a storm cloud and the ground, and which is usually made visible by the presence of a funnel-shaped cloud (a 'tuba') extending from the main cloud base of the storm, and by dust and debris lofted by the tornado. Tornado Alley is a nickname given to part of the central USA that is prone to the development of strong to violent tornadoes. In spring and summer, warm, moist air from the Gulf of Mexico spreads north into the continent, whilst cold air at higher levels spreads from the west over the top of the warm air, resulting in a very unstable atmosphere. Where this instability coincides with winds that exhibit marked changes in speed and direction with height through the unstable layer, severe storms called supercells may form. Strong to violent tornadoes, which in exceptional cases may produce wind speeds up to 483 kph (300 mph), are usually associated with such storms. The strongest tornadoes are capable of catastrophic damage, sometimes not only destroying whole buildings, but actually scouring the surface of the Earth and sweeping the foundation stones of buildings clear of debris, which are then deposited further along the tornado's track.

Anvil of thunder

Mark Mascilli

Lightning ripples through the anvil of an isolated evening supercell thunderstorm near Putnam in Oklahoma, USA. The upper parts of a thundercloud are composed mainly of ice crystals. Strong winds in the upper troposphere (the lowest layer of the Earth's atmosphere in which most of our weather occurs), often draw these parts of the cloud out, forming the characteristic anvil of the cumulonimbus cloud – so-called because its shape is reminiscent of a blacksmith's anvil. When upper-level winds are very strong and the storm is long-lived, the anvil may spread as much as several hundred kilometres ahead of the storm core.

Canon EOS 6D Mark II + EF 24–105 mm *f*/4L IS II USM: 8/5 s; *f*/22; 24 mm; ISO-400.

Lightning shelf

David Vicary

Multiple flashes of cloud-to-cloud lightning illuminate the clouds at the leading edge of a line of high-based thunderstorms near Liberal in Kansas, USA ('high-based' meaning that the main base of the thundercloud is located higher than about 1,500–2,500 m (4,900–8,200 ft) above ground level). David describes what happened, 'After an unsuccessful chase day, we started heading south to Liberal, Kansas. A small storm began to intensify to our west, so we stopped to let it approach us. As the sun fell below the horizon the lightning activity increased, illuminating the structure of the storm clouds, whilst the thickening rain curtains slowly obscured the dwindling twilight to our west.'

Canon EOS 7D Mark II + Sigma 10–20 mm *f*/4-5.6 EX DC HSM: 5/2 s; *f*/10; 20 mm; ISO-400.

Strike at sea

Marc Marco Ripoll

Lightning arcs across the night sky before striking the open sea off the coast of Mallorca, Spain. It is the collision of hail-like particles, known as graupel, with smaller ice crystals in a thunderstorm that results in the build-up of electrical charge within the thundercloud. In the collision process, electrons are stripped from the smaller ice crystals and transferred to the graupel, meaning that the ice crystals end up with a deficit of electrons (meaning they acquire a positive charge) whilst the graupel end up with an excess of electrons (meaning they acquire a negative charge). Due to their differing weights (and therefore speeds of falling) the particles become separated: negatively-charged graupel falls towards the bottom of the cloud whilst the positively-charged ice crystals are lifted by the storm updraft and collect at the top, giving the lower and upper parts of the cloud a negative and positive charge, respectively. Once the charge difference becomes large enough, the insulating property of the air breaks down and lightning (a rapid flow of electrons) occurs in order to neutralize the opposite charges.

Sony ILCE-7M3: 30 s; ISO-500.

Lightning by moonlight

Enric Navarrete Bachs

An almost surreal combination of moonlit skies and a powerful cloud-to-ground lightning bolt is captured looking out to sea from Sant Pol de Mar, Spain. Most cloud-to-ground lightning comprises a flow of electrons to Earth from the negatively charged lower to middle parts of the thundercloud. However, cloud-to-ground lightning bolts occasionally originate much higher in the cloud and involve a net transfer of positive charge from the cloud to the ground. Although relatively uncommon, these positive bolts, sometimes known as 'superbolts', can be extremely powerful and destructive, having a much larger electric current than a typical negative lightning bolt. This is due to the larger air gap that has to be bridged between the upper parts of the cloud and the ground, which thus requires a larger charge to build up before the lightning can occur. The bolt shown here is likely to be an example of a positive cloud-to-ground flash, since it emanates from the upper parts of the cumuliform storm cloud.

Sony ILCE-7M2 + FE 24–70 mm F4 ZA OSS: 8 s; *f*/5; 35 mm; ISO-400.

The Enid supercell

John Finney

This photograph was taken almost directly underneath the updraft of a supercell thunderstorm near Enid in Oklahoma, USA. Turbulent motions result in complex patterns in the cloud base underneath the main updraft, where the cloud base appears particularly dark, and near the interface between the updraft and the downdraft. Shafts of precipitation are visible in the distance, in the downdraft region of the thunderstorm. Supercell thunderstorms are often prolific producers of severe weather, including large hail, tornadoes and damaging non-tornadic wind gusts. Hail larger than golf balls is relatively common in such storms. More rarely, hail may reach the size of softballs or even grapefruits. The largest hailstone ever recorded occurred near Vivian in South Dakota, USA, on 23 July 2010. It measured an astonishing 20 cm (8 in) in diameter and weighed 0.88 kg (2 lbs).

Nikon D810 + 14–24 mm *f*/2.8: 2 s; *f*/11; 14 mm; ISO-64.

Storm in a spin

Arron Hiscox

The setting sun disappears behind an intensifying low-precipitation supercell near Broken Bow in Nebraska, USA. Low-precipitation storms, as their name suggests, are characterized by a relative lack of precipitation, rendering visible much of the storm's structure. Notwithstanding their name, these storms are capable of producing large or even giant hail, which tends to fall sparsely near the interface between the rotating updraft tower and the storm's downdraft region. The updraft tower of this low-precipitation supercell, seen near the centre of the image, appears almost as an isolated mass of cloud, but it is in fact connected to a much larger mass of high-level cloud in the storm's anvil, the edge of which can be seen towards the left-hand side of the image. Prominent striations in the lower reaches of the updraft tower are a sign of strong rotation.

Nikon D5000 + 17–70 mm *f*/2.8–4.0: 1/200 s; *f*/10; 17 mm; ISO-200.

Supercell striations

Arron Hiscox

A high-precipitation supercell thunderstorm dominates the skyline in Nebraska, USA. The prominent cloud bands in this image are forming near the interface between the updraft and the rear flank downdraft of the supercell thunderstorm. The rear flank downdraft is a focused region of strong downdrafts that wraps around the rotating updraft of the storm. In high-precipitation storms such as this one, the rear flank downdraft carries large amounts of precipitation with it and may advance very rapidly. This is a particularly dangerous part of the storm because it is often associated with damaging outflow winds – winds associated with the downdrafts of air which spread out on hitting the ground – in excess of 100–120 kph (62–75 mph) and wind-blown large hail.

Nikon D500 + 16–80 mm *f*/2.8–4.0: 1/20 s; *f*/10; 16 mm; ISO-100.

Lightning on the dancefloor

Boris Jordan

Lightning ripples across the sky in a large thunderstorm to the west of Frankfurt, Germany. Anvil crawler lightning is so called because it occurs within the anvil of the thunderstorm. The anvil is the flattened, upper part of the storm cloud composed mostly of ice crystals. Anvil crawler lightning is often highly branched and it may extend for many kilometres horizontally, especially within larger complexes of thunderstorms. Sometimes this type of lightning is called spider lightning, because the individual branches may look a bit like the legs of a spider, especially where they radiate out from a single point. Recent measurements from satellites suggest that the longest lightning channels may extend as far as several hundred kilometres horizontally.

Canon EOS 6D + Canon EF 50 mm *f*/1.4 USM: 30 s; *f*/7.1; 50 mm; ISO-125.

Windmill under attack

Tina Wright

Inky black storm clouds contrast brilliantly with sunlit fields in the High Plains of northeastern Colorado, USA. Storm clouds appear to be dark when viewed from below because of the great vertical extent and the high water content of the cloud mass. In a typical summer thunderstorm in the middle latitudes, the cloud tops reach around 12–14 km (7½–8½ miles) above ground level. In tropical regions, thunderstorm cloud tops frequently exceed 20 km (12½ miles). The depth of the thunderstorm is limited by the height of the tropopause – a stable layer at the top of the troposphere, which is the lowest layer of the Earth's atmosphere in which most of our weather occurs. The tropopause acts like a lid to rising air and so the cloud top spreads out horizontally underneath it, creating the thunderstorm's characteristic anvil. The tropopause is highest in the tropics, which is why the storms there tend to be the deepest.

Nikon D750 + 28–300 mm *f*/3.5–5.6: 1/60 s; *f*/8; 28 mm; ISO-100.

Early morning strike

Lauren Bailey

A powerful cloud-to-ground lightning bolt strikes within the city limits of El Paso, Texas, USA, in an early morning thunderstorm that had moved north from neighbouring Mexico. High-based thunderstorms – those in which the main cloud base is located well above the surface, usually at a height of 1,500–2,500 m (4,900–8,200 ft) or more, such as the one photographed here – often produce photogenic lightning bolts because much of the lightning channel is visible below the cloud base. Here, the lightning is particularly visible because it also occurs outside the compact core of heavy rainfall associated with the storm, which can be seen over the hills in the distance.

Canon EOS 5D Mark III + 24–70 mm: 1/6 s; *f*/9; 24 mm; ISO-100.

Battling the winds

Fereshteh Eslahi

A group of people struggle to make headway against the strong winds in a blinding dust storm in Bandar Mahshahr, southern Iran. Dust storms, or sandstorms, are a common weather hazard in the more arid regions of the world, arising when strong winds lift particles of dust and sand from the dry ground. The airborne dust may rise to great heights and travel enormous distances. For example, dust from the Sahara is occasionally deposited by rain in parts of northwest Europe, and dust from storms in China is sometimes detected in the USA, after transportation across the entire Pacific Ocean by strong westerly winds in the upper atmosphere.

Canon EOS 6D + EF 24–105 mm *f*/4L IS USM: 1/500 s; *f*/5; 105 mm; ISO-250.

Burning haze

Jacquie Matechuk

Smoke floods the prairies near Cochrane in Alberta, Canada, following weeks of forest fires in the interior of British Columbia. The smoke and suspended particles act to reduce the visibility, giving a persistent haze. Smoke particles preferentially absorb blue and green wavelengths of light and scatter red light strongly, and as a result the sky takes on an abnormal reddish tint. Crepuscular rays can also be seen due to scattering of sunlight by particles suspended in the atmosphere.

Canon 5D Mark IV + Canon EF 100–400 mm II: 1/320 s; *f*/4.5; 150 mm; ISO-800.

Our climate and the natural world

LESLEY HUGHES

Natural ecosystems and the services they provide are the life support system for human existence and our social and economic well-being. The Earth's species provide our food and fibre, reduce the risks of natural hazards, maintain clean air and water, pollinate our crops, store carbon, and provide medicines and pest control. For many, the natural world also provides important spiritual and cultural values. The human fingerprint on this natural support system is vast. We divert over half of all freshwater and have exploited around half of the land surface for agriculture and urban settlements. Humans and their livestock comprise more than 95% of all mammalian biomass on Earth. These interventions have come at great cost to the world's biodiversity and now climate change, particularly the increase in severity and frequency of extreme events, is accelerating the impacts.

Our oceans are absorbing over 90% of the extra heat in the Earth's system, with warming now measurable down to at least 2 km (1¼ miles) depth, and marine ecosystems are feeling the effects. Underwater heatwaves, where bodies of water several degrees above normal stay in place for days or weeks, have doubled in frequency since the 1980s. These heatwaves have wreaked havoc on coral reefs, the most biodiverse marine ecosystems. Corals are symbiotic organisms, relying on microscopic algae in their tissues to photosynthesize and provide food. When stressed by heat, the corals expel these algae, and if the heatwave lasts too long, the corals can bleach and die. Mass bleaching of the world's coral reefs is occurring with increasing frequency and the outlook is equally grim, with predictions of near total loss with low levels of further warming and follow on impacts to all the other species that the reefs support, including the estimated 500 million people who rely on reefs for food, coastal protection and incomes.

Underwater heatwaves are also devastating other marine and coastal ecosystems, including seagrass meadows and mangroves that provide nursery grounds for fish, invertebrates and marine mammals, as well as filtering sediments and storing carbon. But heat is not the only stress on marine systems. As sea level rise accelerates, the impacts of storm surges are exacerbated, damaging corals, kelp forests, and rocky intertidal areas. Many coastal systems are becoming increasingly squeezed between the high tide mark and coastal developments, and saltwater is intruding into coastal rivers, transforming freshwater wetlands to estuarine systems. Absorption of carbon dioxide from the atmosphere is acidifying seawater, increasing the risk that skeletons and shells of marine creatures will simply start to dissolve. Melting polar sea ice is also affecting marine food webs. As the algae that live in nutrient-rich pockets in the ice decline, reverberating effects are being felt right up the food chain.

The increasing severity and frequency of extreme events are also causing enormous damage on land. Wildfires, unprecedented in extent and ferocity, are transforming many terrestrial ecosystems, with vegetation such as rainforests being burned in some regions for the first time in hundreds, perhaps thousands of years. Fires in the Amazon are devastating the 'lungs of the planet', threatening not just thousands of species and indigenous peoples, but potentially transforming the rainforests into open savanna woodland. This could represent a key tipping point in the Earth's climate system, turning the region into a carbon source rather than a sink.

In many regions, heatwaves combined with drought have killed hundreds of thousands of sensitive species like flying foxes and birds. Where rainfall is declining, freshwater ecosystems and species are suffering from reduced flows and poorer water quality, leading to increasing isolation and fragmentation of aquatic systems. In mountainous and polar regions, loss of snow and the rapid melting of glaciers is exposing new ground to be colonized, but also leading to increased erosion, loss of freshwater and hydropower resources, and increasing threats to downstream communities from outburst floods from glacial lakes.

Many species are already adapting to their changed conditions. In mountainous areas, species as diverse as North American pikas and Bornean atlas moths are seeking cooler climates by moving to higher altitudes but eventually may have nowhere to go. Other species, especially in the oceans, are moving to higher latitudes - tropical marine species are establishing in temperate oceans, and temperate species in polar regions, creating both positive and negative impacts on the world's fisheries, affecting the livelihoods and nutrition of the millions of people who rely on marine resources. The rate of these migrations to higher altitudes and latitudes is accelerating, increasing 2–3-fold per decade.

Earlier springs are also advancing life cycle events such as flowering, fruiting and migration. In some cases, these changes are creating mismatches in relationships between species. Insects such as bees, flies, butterflies and beetles in Europe are emerging and flying earlier, but for fewer days, reducing their overlap with the flowers of the plants they pollinate. Warming is even changing the sex ratio of species whose gender is determined by temperature during development. Hatchlings of green turtles, for example, are increasingly developing as females. Some species responding positively to climatic change are creating problems for others, but for every species advantaged by changed conditions, many more are suffering. Species without the mobility to find new habitats are declining, and the rate of species extinction, already 100–1,000 times higher than that in the past, will likely continue to accelerate. A recent global assessment found that around a million animal and plant species are threatened with extinction, many within decades – the sixth mass extinction is demonstrably underway.

As species respond to changing conditions in highly individual ways, the composition of ecological communities is also changing, with novel combinations of species that have likely never coexisted before. Eventually, wholescale shifts in major vegetation communities are expected – the tundra forests at high northern latitudes becoming boreal forests, and woodlands shifting to grasslands in the African Sahel. Human agriculture is being similarly affected. In some regions, warming and rainfall change means that new crops and livestock are now being farmed, whereas in others, farming land is becoming ever more marginal.

Changed distributions of pathogens, parasites and vectors like mosquitoes that carry diseases such as malaria are affecting natural systems, agriculture, fisheries and human health. Other health impacts are being caused by heat extremes. A three-week heatwave in Europe in 2003, for example, is estimated to have killed 70,000 people. Approximately 20 million people per year are already being displaced from their homes and livelihoods by climate-related events, an accelerating global security problem.

Given the transformative change with just over one degree of global warming, the prospect of 2, 3 or 4°C (3.6, 5.4 or 7.2°F) warming can scarcely be imagined. So what can we do to save our life support system? Fixing the cause of the problem is fundamental – we must rapidly slow and then eliminate emissions of carbon dioxide and other greenhouse gases to meet the Paris Climate Agreement goal of limiting warming to well below 2°C (3.6°F). But in the meantime, we need to facilitate natural adaptation where we can, minimizing biodiversity loss and maintaining the services that ecosystems provide. As we adapt to our own changing circumstances, we must ensure that we don't make things worse for biodiversity – building dams and sea walls, cutting fire breaks, clearing land in new regions for agriculture may have even greater impacts on the natural world than the direct impacts of the changed climate. Instead, a focus on nature-based solutions such as living shorelines, restoring riparian vegetation to reduce flooding and erosion, and replanting mangroves and restoring coral reefs to protect shorelines from storms will help both nature and people adapt. It's not too late to act, but what we do over the next few years will be critical not just for the fate of the Earth's biodiversity, but also our own.

Coated in ice

Artur Vinokurov

Ice coats the clothes of a young fisherman as he returns home after braving blizzard conditions and extreme cold in Yakutia, Russia. The poles are the coldest areas on Earth because they receive less energy from the sun than lower latitude regions. Winter cold is also particularly extreme in continental interiors, even in the middle latitudes. This is because these areas are far removed from the moderating influence of seas and oceans. The coldest, permanently inhabited place on Earth is considered to be Oymyakon, situated deep in eastern Russia. In addition to its continental location, extreme cold is also more likely there because Oymyakon is situated within the broad valley of the Indigirka River, where cold, dense air tends to collect and persist. In January 1924, a temperature of -71.2°C (-96.2°F) was recorded at the nearby weather station, an event that is commemorated by a memorial in the village square.

Canon PowerShot SX500: 1/60 s; *f*/4; 4 mm; ISO-100.

A city afloat

Debarshi Mukherjee

A hand-pulled rickshaw struggles through the flooded streets of Kolkata during the monsoon season. Although the Indian Monsoon occurs with a fairly regular seasonal cycle, significant variations can occur in the intensity and timing of the monsoon rains from year to year. Where monsoon rains fail, the results can be catastrophic. Although crops require the monsoon rains in order to grow, excess rainfall can be really problematic, creating widespread flooding that can displace millions of people. The most dangerous flooding events in the region are often associated with landfalling tropical storms in the Bay of Bengal. In May 2020, an exceptionally powerful tropical storm called Cyclone Amphan struck Bangladesh and eastern parts of India, with sustained wind speeds of over 240 kph (150 mph). A storm surge and exceptionally heavy rainfall combined to create widespread flooding. In Kolkata, 240 mm (9 in) of rain was recorded.

Nikon D7000 + Nikkor DX VR 18–55 mm: 1/250 s; *f*/9; ISO-800.

AHCHA

Elegance in the rain

Patrick Hochner

A young maiko (geisha apprentice) protects herself from heavy rain with a traditional 'bangasa', a waxed paper umbrella, while walking in Hanamachi or Flower Town, one of the geisha districts in Kyoto, Japan. Rain is the liquid form of precipitation – it may have started life as an ice crystal which subsequently melted as it fell, or as a liquid droplet. Raindrops can be up to 6 mm (¼ in) in diameter and fall at an average speed of around 22 kph (14 mph). However, larger raindrops can fall as fast as 32 kph (20 mph). Rainfall may be described as being of light, moderate or heavy intensity, as measured by the instantaneous rainfall rate in millimetres per hour. In showery precipitation, heavy rainfall is defined as a precipitation rate between 10 and 50 mm (½ and 2 in) per hour. Very high rainfall rates are seldom sustained for long periods, however, so typically only a few millimetres of rain falls in an individual heavy shower.

Nikon D500 + 70–200 mm *f*/2.8 E FL ED VR: 1/400 s; *f*/2.8; 200 mm; ISO-200.

Dam wet

Andrew McCaren

Water overspills the dam wall at Wet Sleddale Reservoir in Cumbria, England, after a period of heavy rainfall. Climate simulations show that rainfall extremes are likely to increase in a warming climate, which will put increased pressure on infrastructure such as dam walls, water courses and road and rail networks. Between 2000 and 2009, more than 200 significant dam failures occurred worldwide. One of the worst failures on record occurred in China on 8 August 1975, when the Banqiao and Shimantan Dams in Zhumadian Prefecture collapsed following extreme rainfall associated with Typhoon Nina. The resulting flood killed an estimated 171,000 people and displaced approximately 11 million.

Canon EOS 1DX + EF 70–200 mm *f*/2.8L IS II USM: 1/320 s; *f*/5; 200 mm; ISO-400.

Road to nowhere

Mohammad Hossein Moheimani

A road is blocked by swelling floodwaters near the village of Agqqla in northern Iran, following a period of torrential rainfall. Floods are amongst the most costly severe weather events when measured in terms of total economic loss. Combined annual losses from floods are increasing in many countries of the world, driven by increased exposure to the hazard, as infrastructure and housing expands to accommodate increasing populations, and also by the increasing frequency and severity of rainfall extremes in many countries.

DJI FC2103: 1/690 s; *f*/2.8; 4.5 mm; ISO-100.

Sunset rolling in

Anne Elizabeth Mitchell

An extensive layer of altocumulus clouds catches the evening sunlight over the River Arno in Pisa, Italy, making for a spectacular sunset. Altocumulus clouds are medium-level clouds, meaning that they form at altitudes of around 2–6 km (1¼–3¾ miles) in the middle latitudes. They are generally composed of supercooled water droplets, which are liquid water droplets with a temperature below 0°C (32°F). Altocumulus clouds exhibit many different varieties and forms, but they are always characterized by globular masses of rolls of cloud, arranged either in layers or discrete patches. In this case, the altocumulus forms an extensive layer and the individual cloud elements have relatively regular size, so the cloud species is altocumulus stratiformis. This is by far the most commonly occurring type of altocumulus. The individual cloud elements are caused by shallow convection currents or turbulence, for example where a layer of cloud is gently heated from below.

Fujifilm X-E3 + Fujifilm XF 18–55 mm *f*/2.8–4.0 R LM OIS: 1/60 s; *f*/11; 18 mm; ISO-400.

Flash floods in the desert

Guy Nesher

Salar de Uyuni, in southwest Bolivia, is the world's largest salt flat, covering an area of approximately 10,000 square km (4,000 square miles). Up to approximately 12,000 years ago the area was covered by a giant lake, but the water evaporated over time leaving behind a thick crust of salt. In today's climate, the Bolivian Salt Flats are arid, experiencing less than 200 mm (7¾ in) of precipitation a year on average. Whilst deserts are characterized by their general lack of precipitation, occasional heavy rainfall events do occur in many of the world's deserts. These events tend to be brief but intense, and sometimes result in flash-flooding as the water runs off the parched, largely unvegetated ground. In the Bolivian Salt Flats, rainfall occurs most frequently in January. Sometimes, as here, water ponds on the surface of the salt crust after rainstorms, creating a spectacular mirror-like effect.

Canon EOS Digital Rebel XT + 18–55 mm: 1/200 s; *f*/10; 55 mm; ISO-100.

Marching on ice

Yuanyi

A group of swans form an orderly line as they walk across a frozen lake in Pinglu County, Shanxi Province, China. Continental interiors experience much larger seasonal temperature changes than oceanic locations at similar latitudes, because land has a much smaller heat capacity than water and so responds much more rapidly to the seasonal changes in the amount of energy received from the sun. In winter, when there is a net loss of heat, continental interiors cool much more rapidly, whilst in summer, when there is a net input of heat, they warm much more rapidly. So winters tend to be colder and summers hotter in continental regions, when compared to oceanic regions. If you compare monthly average temperatures for Moscow, which has a continental climate, and Edinburgh, which has a maritime climate (both of which are at similar latitudes), this difference is evident. In January, the average daily maximum temperature in Moscow is -5°C (23°F), compared to Edinburgh 7°C (45°F), and in July, Moscow has an average daily maximum of 25°C (77°F), compared to 18°C (64°F), in Edinburgh.

NIKON D810 + 70–300 mm *f*/4.5–5.6: 1/8000 s; *f*/6.3; 195 mm; ISO-640.

No place to hide

Ruediger Schulz

A hare stands out conspicuously amidst the white of the snow-covered landscape in Saxony, Germany. Snowflakes are formed either by the growth of ice crystals within a cloud, or by the collision and aggregation of existing ice crystals. The shape of the flake depends on the temperature, with numerous different forms possible including columns, plates, dendrites, sectored plates and needles. When temperatures are close to freezing point, hexagonal plates and the familiar dendritic forms (complex crystals with six-sized symmetry) are favoured, whereas lower temperatures favour needles, hexagonal prisms and hollow columns. As temperatures warm towards 0°C (32°F), individual flakes become sticky and may aggregate, leading to the formation of larger snowflakes. Exceptionally, the diameter of aggregated flakes may reach several inches.

Canon EOS 5D Mark III + 150–600 mm *f*/5–6.3 DG OS HSM Sports 014: 1/1250 s; *f*/7.1; 600 mm; ISO-400.

Travelling in the rain

Debarchan Chatterjee

People maintain good spirits in spite of the deep flooding in the streets of Kolkata during the Indian Monsoon. Surface water and river flooding are widespread in India during the monsoon season, especially in urban areas where drainage is often poor. In Kolkata, the wettest months are June, July, August and September; rainfall totals average 1,336 mm (52½ in) during this period, amounting to almost three-quarters of the annual average rainfall. The Indian Monsoon is one of the most important atmospheric circulation patterns on Earth, and it may be thought of as a kind of giant sea breeze. During the spring, strong heating occurs over India, which reduces the density of the air and encourages it to rise. This results in the formation of a 'heat low' inland, as the atmospheric pressure falls near ground level. In response, winds begin to flow in from the south and southwest, carrying moist air from the Indian Ocean onto the Indian subcontinent. This moist air progresses further inland as the summer proceeds and it is associated with copious rainfall, especially where the air is forced to rise over higher ground such as at the foothills of the Himalaya. During autumn and winter, the land and overlying air cool and so the atmospheric pressure rises as the density of the air increases. Consequently, there is a reversal of the winds as air begins to flow away from the high pressure region, such that relatively dry north-easterly winds originating over the land prevail. This annual cycle explains the strong seasonality of rainfall in regions affected by the monsoon.

Nikon D3200 + 18–55 mm *f*/3.5–5.6: 1/500 s; *f*/5.6; 18 mm; ISO-1600.

78/1 BABUGHAT
TIMES NOW

Framed by a rainbow

Joann Randles

A rain-soaked, semi-feral Welsh mountain pony grazes on Cefn Bryn Common near Swansea, Wales, in the wake of a heavy hailstorm. Rain accentuates the effects of cold because water conducts heat away from the body more effectively than air. Many wild animals are rather better adapted to this problem than humans because of their thick coats of fur. The individual hairs act to trap warm air close to the body, therefore insulating the animal from the effects of the cold. Ponies such as this are often used for conservation grazing schemes.

Canon 5D Mark IV + Canon 16–35 mm: 1/1600 s; *f*/2.8; 16 mm; ISO-320.

The smell of rain

Charley Nicholls

A young vixen weathers a sudden spring hailstorm in Rayleigh, Essex, England. Before it begins to rain, the humidity of the air tends to increase. This causes the pores of rocks and soil to fill with moisture, forcing some of the oils from the earth into the air, releasing fragrant chemical compounds at the same time. However, the strongest scent generally occurs when the rainfall arrives. As a raindrop hits the ground, small air bubbles become trapped on the surface, which then shoot upwards and burst out of the drop, throwing aerosols into the air where they are then distributed by the wind. The main contributor to the resulting scent is a compound called geosmin that is released by actinobacteria – tiny microorganisms that work to decompose organic matter. The smell – which is called petrichor – is often most noticeable just as the rain begins, especially when the ground is relatively warm.

Canon EOS 7D Mark II + 150–600 mm *f*/5–6.3 DG OS HSM Contemporary 015: 1/100 s; *f*/8; 600 mm; ISO-640.

Our climate and Earth's frozen water

MICHALEA KING

Climate change has impacted all major components of the environment, but some of the most dramatic changes are being observed within the frozen parts of the Earth, known collectively as the cryosphere. This includes snow and ice in all its various forms: ice caps and mountain glaciers, sea, lake and river ice and frozen ground or permafrost. The cryosphere is especially sensitive to climate change because so much of Earth's ice is located in regions that are warming faster than the global average. This includes snow and ice in rapidly warming areas like the polar regions, for example the Arctic and the West Antarctic Peninsula, and glaciers in mountainous regions around the world. Some components of the cryosphere, such as some seasonal sea ice and snow cover, melt and reform every year, while others have been features of their landscapes for millennia. Climate change has affected nearly all aspects of the cryosphere to some degree, and these changes in turn initiate other processes, such as rising sea levels, changing air and ocean circulation patterns, and an increase in the amount of sunlight absorbed on the Earth's surface.

Scientists have observed reductions in the depth and extent of seasonal snow cover, as well as declines in the duration of snow cover over much of the Arctic. These decreasing trends in depth and coverage are also mirrored in permafrost, or the icy layers of the ground that remain frozen year-round. The widespread thawing of permafrost has serious ecological and societal implications, such as destabilizing existing infrastructure, the contamination of groundwater, and the potential release of stored greenhouse gases from the soil.

Of all the seasonally changing components of the cryosphere, however, some of the most striking changes are occurring in sea ice. This feature of the cryosphere forms as an icy crust on the surface of ocean waters that have cooled to the point of freezing. In the Arctic, sea ice extends southward from the central Arctic during the dark winter months. With the return of sunlight to the Arctic, sea ice then recedes back to within the central Arctic Ocean during the warmer months, typically reaching a minimum extent in September. Detailed satellite observations, however, have shown a rapid reduction in ice thickness and extent, with the minimum September extent decreasing by over 40%, and ice thickness by about 50% since records began in 1979. Although melting sea ice does not contribute to sea level rise because the ice is already floating on water, sea ice is still very important because it helps keep the Arctic cool by reflecting most of the incoming sunlight away from Earth's surface. As Arctic sea ice continues to retreat, greater areas of dark ocean surface are exposed. Unlike sea ice, which is bright and highly reflective, the dark underlying ocean absorbs most of the incoming solar radiation, amplifying the existing warming trend. Current rates of change point to an increasing likelihood of ice-free summers in the Arctic by the end of the century, and as early as 2050 under high greenhouse gas emission scenarios. Conversely, no significant trends have been observed in Antarctic sea ice, though the configuration of sea ice there is fundamentally different to the ice forming in the Arctic Ocean. In Antarctica, sea ice instead forms at the edge of the continent and extends northward into the Southern Ocean during the winter months, effectively doubling the size of the continent. Nearly all of this seasonal ice retreats back to the edge of Antarctic in summer, with very little ice surviving into the following year. Because so little summertime sea ice exists around Antarctica, changes have less of an impact on how much solar radiation is absorbed at the ocean surface, unlike the Arctic Ocean.

The bulk of the Earth's freshwater (over 99%) is stored in solid form within two massive polar ice sheets. Ice sheets are massive bodies of land-based ice that have formed from the gradual layering of snowfall that compressed into ice over thousands of years. The melting of these two ice sheets is currently raising sea levels at a rate of approximately 1 to 1.5 mm each year. The southernmost continent of Antarctica is covered nearly entirely by the larger Antarctic Ice Sheet, which can be as thick as nearly 5 km (3 miles) in some areas and altogether holds enough ice to raise sea levels by 58 m (190 ft). The Greenland Ice Sheet, located in the northern Atlantic Ocean, is much smaller by comparison, but still stores enough ice to raise sea levels by 7 m (23 ft) if melted in its entirety. Because of their immense thickness, the surfaces of ice sheets are located at high elevations and are therefore very cold. However, in the warmer summer months, temperatures can rise above freezing, causing some of the snow and ice at the surface to melt. Warmer air temperatures, resulting in longer and more intense melt seasons, have increased the volume and areal extent of surface melt both over the Greenland Ice Sheet, as well as over large areas of Antarctic ice shelves, which are semi-permanent floating slabs of ice that form when ice flows off the ice sheet over cold ocean water. Melting at the surface is a major contributor to the overall accelerated rates of ice loss in Greenland, which recorded a record-breaking annual loss of nearly 600 Gt (1 Gt = 1 billion metric tons) in 2019. Antarctica is also losing mass, but only about half as rapidly as Greenland, with most of its mass loss being concentrated in the West Antarctic Ice Sheet and Antarctic Peninsula, a region which holds the ice equivalent of over 3 m (10 ft) of sea level rise. Although more ice is currently being lost from Greenland, much of the West Antarctic Ice Sheet rests on a bed below sea level and is therefore more vulnerable to rapid melt, resulting in a greater potential for large ice losses from Antarctica in the coming decades to centuries.

Enhanced surface melt is not the only process responsible for shrinking ice sheets. At the margins of the ice sheets, the ice flows out toward the ocean as frozen rivers of ice called glaciers. Many of these large glaciers flow down to the ocean through deep fjords, with their beds resting below sea level. Here, ice at the front of glaciers either melts from direct contact with water, or calves off in pieces, forming icebergs that drift off to sea. These processes transport previously land-based ice and water into the ocean, and thus contribute to sea-level rise. This is different from the loss of floating ice such as ice shelves and the retreat of sea ice, which will not displace additional sea water once melted. Many massive glaciers in Greenland began flowing faster and draining more ice to the ocean around the year 2000. Now, more ice is consistently lost through the flow of glaciers along the edge than can be regained through snow accumulation on the ice sheet's surface. In Antarctica, the disintegration and thinning of ice shelves, which act as a plug to help buttress thick interior ice, threatens to accelerate ice loss from the ice sheet in the future. So far, rates of mass loss from ice sheets have followed the upper range of modeling projections, which reflect the rates of ice loss expected with unmitigated emissions of greenhouse gas emissions.

Glaciers are not only found along the edges of Earth's two ice sheets, but also form in mountainous regions around the world, which are home to approximately 10% of the global human population. Changes in mountainous glacial environments directly impact humans and ecosystems that depend on the timing and magnitude of glacial meltwater runoff to survive. Warming air temperatures are causing rapid reductions in glacier volume and area, as well as causing shifts in the timing of availability of meltwater from alpine glaciers to downstream valleys. Concerningly, scientists estimate that under high greenhouse gas emission scenarios, many alpine ranges home to smaller glaciers will see dramatic reductions in ice volume, including a loss of up to about 90% of mountainous glaciers in the European Alps by 2100. Though small in size compared to the massive ice sheets, the rapid reduction of alpine glaciers threatens both the environment and society in ways that extend beyond the ever-present concern of sea-level rise. Shrinking alpine glaciers join other facets of the cryosphere in their reduced extent and volume, with recent work estimating that combined, climate change has resulted in a loss of nearly 30 trillion tons of ice worldwide since 1994.

As the oceans and atmosphere continue to warm, the cryosphere will become increasingly at risk, requiring concerted global efforts to mitigate current trends and preserve remaining ice.

Aurora at night

Serhii Kurlia

The aurora borealis appears over a snow-covered Kirkjufell on the west coast of Iceland. Aurora borealis, also known as the northern lights, occurs when energetic charged particles from the sun are deflected by the Earth's permanent magnetic field and interact with gases in the upper atmosphere. This usually occurs at high latitudes within the 'auroral oval' – a huge ring centred on the Earth's magnetic north pole – but it may extend to lower latitudes in geomagnetic storms. Serhii describes the experience, 'A friend and I travelled to Kirkjufell with the aim of capturing this spectacular isolated mountain with the aurora borealis behind. After two nights with stubborn cloud cover our chances were running out. Fortunately, the weather improved during the early hours of our last night, allowing us finally to capture the shot that we had been hoping for.'

Olympus E-M1 Mark II + Olympus M.8 mm *f*/1.8: 30 s; *f*/1.8; 8 mm; ISO-400.

Teeth of ice

Phil Koch

Extreme cold and wind-blown spray combine to create a spectacular ice-bound landscape on the shores of Lake Michigan in Wisconsin, USA. For ice formations to build up like this, a substantial period of sub-zero temperatures is required in order for the surface of the lake to freeze. The time lag between the onset of sub-zero air temperatures and the freezing of the lake is due to the large heat capacity of water – a great deal of energy must be lost to the surroundings in order for the temperature of the lake to decrease appreciably, and so the lake cools much more slowly than the surrounding land surface and overlying air. Early in the winter, when the lake is still ice-free, but air temperatures are already well below freezing, waves crashing onto the shore create spray that freezes quickly on contact with objects near the shoreline.

Canon EOS 7D: 1/40 s; *f*/16; 17 mm; ISO-320.

The smoking mountains

Marek Kosiba

Cumulus and stratocumulus clouds shroud the summits of the volcanoes Licancabur (left) and Juriques (right) in the Andes, on the southwestern Bolivian border with Chile. Cumulus clouds are so-called because they comprise well-defined, rounded masses that, in the right conditions, build up to form large turrets with cauliflower-like tops ('cumulus' means 'heap' or 'pile' in Latin). They form due to convection within the atmosphere. When the sun heats the ground, the air immediately above it is warmed in turn. Local variations in the amount of heating due, for example, to differences in the vegetation cover, soil moisture or topography lead to the development of pockets of air that are warmer than their surroundings. Since warm air is less dense than cold air, these warm pockets begin to rise. The rising air cools and expands and, with enough cooling, moisture condenses out to form water droplets and a cumulus cloud is born. Sometimes, if the rising currents of air encounter a stable layer higher in the atmosphere, the individual cumulus clouds begin to spread out and eventually they merge, resulting in the development of stratocumulus – 'strato' meaning 'layer' in Latin. There are two main reasons why snow depth tends to increase with altitude. Firstly, the temperature decreases with altitude, so precipitation is more likely to fall as snow in the first place, whilst any lying snow tends to persist for longer. Secondly, precipitation is more frequent and often heavier over hills and mountains, due to the forced lifting of air so, where precipitation falls as snow, the accumulations are larger over the mountains.

Canon PowerShot Pro1: 1/1250 s; *f*/4; 7.188 mm; ISO-50.

Sparkling jewels of ice

Alexey Trofimov

Blocks of ice on the frozen surface of Lake Baikal, Russia, appear to shine like precious stones beneath a deep layer of powdery snow. Ice takes on a blue hue because the longer, red wavelengths are absorbed whilst the shorter, blue wavelengths are transmitted and scattered within the ice. The longer the path of the light through the ice, the more blue it appears. Lake Baikal, located in southern Siberia, is the largest freshwater lake in the world by volume. It is ice-covered for several months in the winter. Ice blocks such as these may form when strong winds break the surface ice up and cause it to accumulate close to the windward shore of the lake.

Canon EOS 500D + Tamron SP AF 10–24 mm *f*/3.5-4.5 Di II LD Aspherical (IF) (B001) Canon EF-S:1/250 s; *f*/11; 10 mm; ISO-100.

Shades of blue

Daniel Sambraus

Small icebergs litter the black sands at Diamond Beach in Iceland. Originating from the retreating glaciers of Vatnajökull National Park, they are carried out to sea before washing onto the beach, their blue-white coloration contrasting with the black volcanic sands. Water has the unusual property of being denser in its liquid state than in its solid state. This explains why icebergs float and may travel for long distances after calving from an ice sheet, before melting or being washed ashore. Because the density differences between liquid water and ice are only small, around 90% of a floating iceberg is below the water's surface, hence the expression 'tip of the iceberg'.

Canon G1X: 8 s; *f*/13; 30.3 mm; ISO-100.

Cactuses in the snow

Tina Wright

'New Year's Day 2019 brought very cold air and a potent storm system to Arizona, with snow levels dropping to less than 600 m (2,000 ft) above sea level. By daybreak the Sonoran Desert was coated in a picturesque layer of snow, something that is rarely seen in this area. This image was taken just east of the city of Phoenix in an area known as the Tonto Basin. The snow was several inches deep in some places and coated the tops of the giant saguaro cacti (*Carnegiea gigantea*). These lowland deserts only see snow once every 5–10 years, on average, so this was a real treat and a wonderful way to bring in the New Year!'

Nikon D850 + 24-120 mm *f*/4.0: 1/20 s; *f*/10; 24 mm; ISO-100.

Fortress of ice

Christoph Schaarschmidt

Snow and rime icing in the Erzgebirge (Ore Mountains) in Saxony, eastern Germany. A strong wind combined with 120 cm (50 in) of fresh snow at -10°C (14°F) created these incredible snow sculptures. Rime icing occurs when fog combines with sub-zero temperatures. The thickest deposits tend to occur on mountain tops in the winter, where sub-zero temperatures persist for weeks or even months at a time, and the cloud base is often below the summits. Very large accumulations of rime ice can build up on surfaces exposed to the wind in these conditions, with deposits sometimes exceeding a metre in thickness. The Fichtelberg (Fichtel Mountain) is often hidden by thick fog – in fact in 1951 the mountain experienced fog on 315 days of the year.

Canon EOS 6D Mark II + 14-24 mm *f*/2.8 DG HSM Art: 3/5 s; *f*/14; 21 mm; ISO-100.

Frost in close up

Dianne Kaye Galbraith

The individual ice crystals comprising hoar frost are shown to beautiful effect in this close-up captured in Kangarilla, Australia. Water is made up of hydrogen and oxygen atoms bonded together; as water freezes this bonding creates an hexagonal crystal lattice, which is reflected in the hexagonal form of many ice crystals. This also explains why snowflakes have six-sided symmetry, since the more complex structures in snowflakes grow from this same basic hexagonal crystal structure. Although snow and frost appear white, the individual crystals are in fact transparent. It is the diffuse reflection of light in many directions from the various faces of the complex crystals that gives snow and frost their white appearance. In this case the ice crystals comprising the hoar frost take the form of hexagonal prisms, so the hexagonal shape is best seen at the ends of the elongated crystals.

Canon EOS 5D Mark II + MP-E 65mm *f*/2.8 1-5x Macro Photo: 1/200 s; *f*/6.3; 65 mm; ISO-100.

Suspended in ice

Kolesnik Stephanie

Air is trapped in ice in Shakhty, Rostov Oblast, Russia. The solubility of air in liquid water is related to temperature: the cooler the water, the more air that can be dissolved within it. However, this situation changes when the water freezes, as gas solubility in ice is much less than in water. Upon freezing, some of the dissolved air therefore separates from the water, in a similar manner to the formation of bubbles of carbon dioxide in fizzy drinks. Some of this air becomes trapped within the ice as it forms. Typically, the quicker the freezing rate, the larger the number and size of the bubbles.

Sony DSC-W90: 1/125 s; *f*/8; 5.8 mm; ISO-100.

Frost and ice

Julie Fryer

Hoar frost and soft rime ice combine to create a wintry scene near Wigton in Cumbria, England. Air frost occurs when the overlying air temperature is also below freezing point. Meteorologists differentiate between ground frost and air frost, both of which result in the deposition of ice crystals at ground level when there is sufficient moisture in the air. Ground frost occurs when the air temperature is a little above freezing point but the ground surface is below freezing, which is possible because the ground tends to become colder than the overlying air as it loses heat efficiently by radiation. Air frost occurs when the overlying air temperature is also below freezing point. The thickest deposits of frost, however, tend to involve soft rime ice. This occurs in freezing fog when the supercooled water droplets comprising the fog freeze on contact with objects exposed to the air, as distinct from the direct deposition from water vapour to ice in the case of hoar frost. Julie says, 'After a night of freezing fog and exceptionally low temperatures, the fog started to lift, the sun came out, and this was the result in the fields near my home. I had to get up early to capture this scene, because I knew that the sun would soon melt the main subject of the photo. The light was incredible, with the blue sky contrasting sharply with the frosty landscape.'

6 x 6 cm twin lens reflex camera: 1/45 s; *f*/11; ISO-50.

Battling through the blizzard

Rudolf Sulgan

Wind-blown heavy snow envelops the Brooklyn Bridge in New York, USA. In the USA, the National Weather Service defines a blizzard as a storm with wind speeds greater than 56 kph (35 mph) and enough snow to reduce visibilities below ½ km (¼ mile) for at least 3 hours. Snow does not always need to be falling for a blizzard to occur – strong winds can pick up snow that has fallen previously, creating a ground blizzard. Conditions can be life-threatening, especially during a 'whiteout' where the visibility is so reduced that it obscures the horizon and even nearby objects, leading to a risk of disorientation for anybody caught in the open. The northeast of the USA is susceptible to occasional severe blizzards, called nor'easters, that occur when a low pressure area deepens close to the eastern seaboard. These storms feed off the intense temperature contrasts between cold air originating in the continental interior of North America and relatively warm air over the North Atlantic Ocean.

Canon EOS 5D Mark III + Canon EF 24–105 mm *f*/4L USM: 1/60 s; *f*/7.1; 75 mm; ISO-1000.

Beach of snow

Shaun Mills

An orderly line of beach huts overlooks the snow-covered beach at Mersea Island, Essex, England. When very cold air masses pass over comparatively warm waters they become unstable, resulting in heavy snow showers near windward coasts. Occasionally, persistent shower bands, or 'streamers', form which are capable of producing large accumulations of snow when they move ashore. These shower bands sometimes affect east and southeast England when winds blow from the east or northeast. However, the best-known example of this phenomenon is probably the 'lake effect' shower bands that occur in cold air outbreaks over the waters of the Great Lakes in the northeast USA. In exceptional cases, these bands can produce several feet of snow in only a few hours; however, their narrowness means that the areas affected are often extremely localized – sometimes only a few miles wide – with neighbouring regions receiving little or no snow.

Canon EOS 5D Mark IV + EF 70–200 mm *f*/2.8L IS II USM: 1/160 s; *f*/4; 110 mm; ISO-100.

Clouds stacked up

Bingyin Sun

Altocumulus lenticularis clouds often exhibit a spectacular multi-layered structure, best seen around sunrise or sunset when the low-angle sunlight illuminates the cloud bases, as in this beautiful example over the glacial lake Jökulsárlón, in Vatnajökull National Park, Iceland. When the cloud is composed of numerous layers stacked one on top of another, the formation is called 'pile d'assiettes', which translates as 'stack of plates'. These clouds form in stable conditions when strong winds flow across mountain barriers. The air is forced upwards as it meets the mountain and then descends to the lee of the mountains. However, the stable conditions mean that the air may continue to oscillate up and down after the initial displacement over the mountains, creating a series of waves that may extend many tens or even hundreds of kilometres downwind of the mountains.

Canon EOS 5D Mark IV + EF 24–70 mm *f*/2.8L USM: 1/13 s; *f*/11; 42 mm; ISO-400.

Feathers of ice

Rogier Nieuwendijk

This macro shot reveals the intricate detail of rime ice formations on a piece of wooden fencing in Baraque Michel (National Park De Hoge Venen), Belgium. A considerable period of sub-zero temperatures and freezing fog had occurred prior to the photograph being taken, during which the rime ice built up on the wood. Since rime ice accumulates on the windward side of objects, the build up of ice creates a record of the wind direction, including the micro-scale variations in the flow created where the wind encounters obstacles, such as the fence panel in this photograph. The resulting effect is an abstract and complex pattern of delicate ice structures, appearing almost like the feathers of a bird.

Nikon D500 + 90 mm *f*/2.8: 1/640 s; *f*/11; 90 mm; ISO-500.

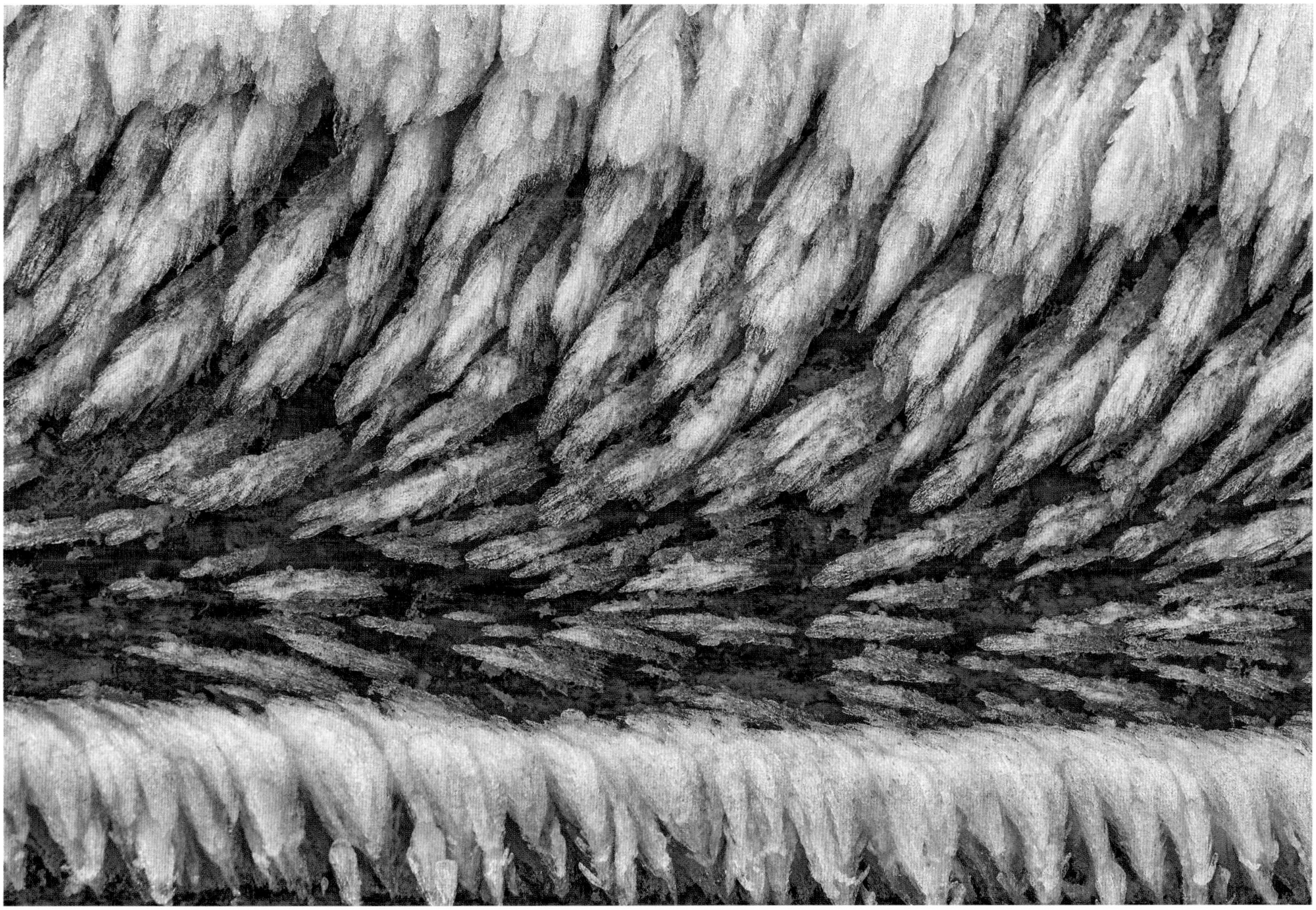

Arctic smoke

Andy Fowlie

A spectacular example of steam fog is captured at dawn close to Suomenlinna Sea Fortress near Helsinki, Finland. Steam fog, also known as arctic smoke, occurs when very cold air flows over comparatively warm waters, such that there is a very large temperature difference between the water and the overlying air. Water vapour is released into the air just above the water which, on mixing with the much colder air above, rapidly condenses to form the fog. With further mixing the fog tends to evaporate, so the fog usually forms a dense but relatively shallow layer immediately above the water. This example occurred when temperatures fell rapidly from around 0°C (32°F) the previous evening to -27°C (-17°F) at the time that the photograph was taken.

Sony ILCE-7RM2 + FE 55 mm *f*/1.8 ZA: 1/8 s; *f*/22; 55 mm; ISO-100.

The snow line

Jason Hudson

A typical winter's scene in the English Lake District, with the snow-capped Langdale Fells contrasting with the lush green of the snow-free fields in the valley below. The deepest level, meaning un-drifted, snow depth on record in an inhabited area of the UK is 1.65 m (5½ ft), as measured near Ruthin, North Wales, in March 1947. However, it is likely that substantially deeper falls have occurred on mountain tops in the UK from time to time, where measurements are generally not available and where drifting can make it nearly impossible to estimate the level snow depths. Globally, the snowiest region is thought to be the Japanese Alps, Honshu Island. At altitudes exceeding about 600 m (1,970 ft) above sea level, average annual snowfall accumulations in this region range from 30–38 m (98–125 ft). On 14 February 1927 the measured snow depth at Mt Ibuki was an incredible 11.8 m (39 ft).

Sony ILCE-7RM2 + FE 70–300 mm *f*/4.5–5.6 G OSS: 1/320 s; *f*/7.1; 117 mm; ISO-100.

Capping the Andes

Francisco Javier Negroni Rodriguez

A striking lenticular cloud hovers over El Chaltén in the Los Glaciares National Park in Patagonia, Argentina. Altocumulus lenticularis clouds are particularly common at the southern tip of the Andes. This is partly because there are very few other land areas and mountain barriers at similar latitudes in the southern hemisphere, and so westerly winds are able to flow almost uninterrupted around the globe across the Southern Ocean. These westerly winds form a well-defined circumpolar current that is strongest in the winter months. After travelling great distances over the ocean, the air within this westerly current becomes moist, but because the ocean is relatively cold, the air also tends to be stable, which inhibits convection. As the moist, stable air is forced to rise on encountering the Andes, lenticular clouds often result.

Nikon D600 + 80–200 mm *f*/2.8: 1/1600 s; *f*/6.3; 86 mm; ISO-100.

A smattering of snow

Alan Leightley

A covering of snow coats Bamburgh Castle in Northumberland, England. Snow showers originating over the North Sea particularly affect the exposed north- and east-facing coastal parts of the UK when winds blow from a north or north-easterly direction. The exact wind direction is critical in determining how far inland these showers spread. When the wind is roughly parallel to the coast, showers are largely restricted to the immediate coastal strip, with locations only a few miles inland experiencing comparatively little or no snowfall. On the other hand, when the wind has a strong onshore component, the showers may be carried much further inland. Alan says, 'After one of the many snow showers that passed through my part of Northumberland this day, I decided to head to the beach so that I could capture Bamburgh Castle in the snow. This required only a two-minute drive and the shot was taken with a tripod from the parking bay. Keeping my fingers warm whilst capturing the photo proved to be the most challenging aspect!'

Sony ILCE-7RM2 + FE 24–70 mm F4 ZA OSS: 1/60 s; *f*/11; 51 mm; ISO-100.

Trees of white

Preston Stoll

Rime-covered trees stand isolated from the surrounding landscape in Medicine Bow-Routt National Forest, Colorado, USA, as a bank of fog rolls in. Rime ice forms in freezing fog, which is the name given to fog occurring in combination with temperatures at or below freezing point. The freezing fog contains supercooled water droplets – that is, liquid water drops at a temperature below freezing point. As soon as these supercooled droplets touch a solid object, such as the trees photographed here, they freeze. Over time, an accumulation of ice builds up on any objects exposed to the fog. The longer the freezing fog persists, the thicker these accumulations become.

Nikon D750 + Nikon 70–200 mm *f*/4: 1/1250 s; *f*/5.6; 200 mm; ISO-100.

Snow roller

Brian Bayliss

Conditions have to be just right for snow rollers to occur, which probably explains why they are so rare. A smooth, un-vegetated hillside, such as in this case near Marlborough, England, increases the chance of their formation. A thin layer of relatively wet snow, settled atop an existing layer of ice or powdery snow, combined with temperatures near the melting point of ice and wind speeds strong enough to move the snow, are all fundamental to the creation of these natural oddities.

Apple iPhone 8 Plus: 1/6024 s; *f*/1.8; 3.99 mm; ISO-20.

Sculptures of ice

Allan Macdougall

Describing the weather he experienced Allan says, 'Pumlumon is a beautiful hill massif in northern Ceredigion, Wales, reaching an altitude of 752 m (2,467 ft) above sea level. I walk this area regularly in all seasons, but winter is my favourite time of year here. The summit plateau is transformed in severe winter weather, presenting challenges even for well-equipped and experienced walkers. In this case, days of wind-blown snow and spindrift from powerful and freezing north-easterly winds had accumulated on every windward vertical surface, combining with the rime ice to form bizarrely shaped natural sculptures on wire fencing and a stile at the top of the ridge.'

Panasonic LX3 + Leica Vario-Summicron: 1/400 s; *f*/11; 12 mm; ISO-100.

Aurora over the forest

Andrei Baskevich

Taking its name from the Latin word for dawn, an aurora occurs when charged particles from the sun are deflected by the Earth's permanent magnetic field and interact with gases in the upper atmosphere, creating visible light. Andrei says, 'Capturing this beautiful scene on Volosianaia Sopkain Mountain in Paanayarvi National Park, Russia, was made difficult by the icy winds, severe frost and waist-deep snow. However, the combination of pristine, deep snow cover and the beautiful aurora made for a striking sight that in the end was well worth the effort. A green curtain of aurora gradually developed and formed into a shape that, with some repositioning, could be made to appear as if it were curving around the snow-covered pine tree, creating this unique composition.'

Nikon D750 + Sigma 20 mm *f*/1.4: 10 s; *f*/7.1, 20 mm; ISO-1000.

Flying saucers

Iain Afshar

A spectacular example of an altocumulus lenticularis cloud is captured over Aiguille du Chardonnet in the French Alps, near Chamonix. These clouds, with their saucer-like shape, have sometimes been mistaken for UFOs and are one of the most common explanations for UFO sightings. As the wind blows across hilly or mountainous regions, air is forced to rise. However, when the air is stable it descends again downwind of the mountains, resulting in a well-defined wave formation. When the air is sufficiently moist, a lenticular cloud forms within the crest of the wave, over or just to the lee of the mountains. Although the wave and the associated lenticular cloud are basically stationary, the air flows continually through them so that the lenticular cloud tends to change shape gradually as it hovers above the mountains, due to small variations in the humidity of the air flowing through, or subtle changes in the wind speed and direction.

Sony P120: 1/640 s, *f*/6.3, 24 mm, ISO-100.

Our climate and the ocean

KATHERINE HUTCHINSON

Humankind has been enriched by the ocean since time immemorial, with the sea providing a platform for trade and transport, a source of food and medicine and a recreational playground. Yet the most valuable service that the sea provides is often overlooked, namely the regulation of global climate. Covering over 70% of the Earth's surface, the ocean plays a major role in the Earth's climate by redistributing heat around our planet. As the Earth is a globe, the angle of incidence of the sun's radiation at high latitudes (the poles) is more oblique than at low latitudes (the equator and tropics). This means that the Arctic and Antarctic receive less intense solar radiation than the tropics and are thus comparatively cooler, explaining the presence of the ice caps. Given this imbalance in heating, one might ask why the equator does not become increasingly warmer and the poles perpetually cooler. The answer lies in the circulation of ocean and atmospheric currents.

A series of ocean currents transport warm equatorial waters towards the North and South Poles, and carry cooler water back again in what is known as the global ocean conveyor belt, or in scientific terms the 'meridional overturning circulation'. Meridional means north-south movement, and overturning means vertical displacement. The timescale of these motions is in the order of decades to centuries, which is why they are so difficult to detect with the naked eye. Superimposed on the daily experience of waves and tides are hundreds of currents moving in different directions at various depths, connecting our planet in a global system.

Our oceans are consequently in constant motion, with bodies of water circulating vertically, latitudinally and longitudinally, carrying heat, salt, oxygen, carbon and nutrients around our planet, sustaining ecosystems and moderating Earth's climate.

For centuries, the ocean was regarded as being immutable. It was considered impossible that human actions could alter the properties or characteristics of this immense body of water. However, with the industrial revolution came increased emissions of greenhouse gases and the start of the greatest geophysical 'experiment' in the history of mankind.

Greenhouse gases trap heat in the Earth's atmosphere in what is known as the greenhouse effect. This process allows for a habitable planet, but an excess of these gases such as carbon dioxide means excessive heating. Since the industrial revolution, humans have continued to increase the amount of greenhouse gas emissions into the atmosphere resulting in an anthropogenic (human induced) global warming of just over 1°C (1.8°F). While the magnitude of this global average temperature change may not be perceptible to the human body, the implications of excess carbon dioxide and extra heating is drastic for both natural and human systems.

The ocean has a great capacity to absorb both carbon dioxide and atmospheric heat. In doing so the sea has borne the brunt of climate change by absorbing an astonishing 93% of the excess heat in the Earth's system and approximately 31% of carbon dioxide emissions. But this service has come at a great cost to ocean health and stability.

Our oceans have persistently warmed over the past 40 years, with even the remotest parts of the sea, the deep abyss and the Southern Ocean, experiencing measurable warming. A rise in ocean temperature has resulted in a doubling in frequency of marine heatwaves, an increase in stratification, a decrease in mixing and a likely weakening of the ocean conveyor belt's overturning circulation.

The climatic consequences of a warmer ocean manifest in more intense tropical cyclones and extreme weather events. Furthermore, changes in sea surface temperature could shift prevailing wind belts which in turn leads to changes in ocean currents. Of concern is how the El Niño Southern Oscillation (ENSO) will be affected by these changes. ENSO is an ocean-climate phenomena that affects atmospheric circulation, temperature and rainfall across the planet through an irregular periodic variation in winds and sea surface temperatures over the tropical eastern Pacific Ocean. Scientists are still exploring the details of how ENSO will operate under global warming, but what is predicted are more severe ENSO related floods, droughts and forest fires, and forced transitions between different states of marine ecosystems.

An interesting characteristic of warmer water is that it takes up more space due to thermal expansion, and so, a warmer ocean means a rise in sea level. Additionally, a warmer ocean and atmosphere leads to a melting of ice sheets and glaciers, which in recent years has overshadowed the effects of thermal expansion by simply adding more water to the ocean. By 2100, the global mean sea level is predicted to rise between 0.43 and 0.84 m (1.41 and 2.76 ft) depending on future emissions, and is projected to continue to rise in subsequent centuries. The impact on low-lying coastal and island communities, about 10% of the world's population, is likely to be increasingly disruptive and, on occasion, catastrophic.

Warming of surface waters has meant increased stratification and less vertical mixing which has affected the ocean's ability to ventilate, i.e. to breathe. The upper ocean has consequently lost oxygen during the last 40 years. Increased absorption of atmospheric carbon dioxide has resulted in ocean acidification which, together with the oxygen loss, has negatively impacted the structure and functioning of the world's most productive large marine ecosystems.

In response to changes in the ocean environment, marine species are having to adapt to new habitats. Under all emission scenarios, a shift in species composition reaching from the sea surface to the seafloor is expected, impacting fisheries and biodiversity. The sinking of organic matter to the ocean floor is projected to decline, implying a decrease in biomass of benthic (bottom dwelling) species. This, combined with ocean warming, oxygen loss and acidification, is expected to drastically harm corals, which have high levels of biodiversity. Where alterations in temperature and oxygen are outside species' tolerance ranges, systems will not be able to overcome the pressures of ocean change, resulting in a loss of animal biomass and the extinction of some species.

The list of ocean changes, both already experienced and projected, is extensive and may be viewed by many as depressing. It is not surprising that many would rather deny that it is underway or underplay the severity of the situation. Hope is, however, not lost. By approaching this problem as a united global community, we are currently in the process of addressing the challenge of monitoring, understanding and mitigating ocean change.

The ocean is a connected system and so this is the approach that we follow in science. Scientists confront the challenge of analyzing and forecasting the ocean's response to climate change on a daily basis. They do this not only to satisfy their curiosity of the natural system but to provide meaningful data and solutions that may be adopted by governments, the private sector and society as a whole. This global approach is necessary as it bypasses national political agendas, allowing scientists to reach across economic and cultural borders to collaborate and share knowledge. The solution is not to build walls and restrict access, but to move past boundaries and aim for open access. Governments too have started to see the necessity of operating as a united community, with the signing of the United Nations' Paris Agreement in 2016 which aims to strengthen the international response to the threat of climate change and limit global warming to a maximum of 2°C (3.6°F), and preferably to 1.5°C (2.7°F). This climate treaty, together with various other agreements on marine reserves, waste reduction and overfishing, has brought the dawn of an age where conserving natural resources takes precedence over exploiting them at any cost.

Much like the ocean reaches across our planet to connect us all, so must we co-operate, communicate and collaborate to work towards a sustainable relationship with the ocean, and hence our climate.

Stacks of gold

Richard Fox

Primary and secondary rainbows appear above sea stacks at Mangersta on the Isle of Lewis, Scotland. Most rainbows last only a few minutes from the perspective of an observer at a fixed location, but occasionally rainbows have been observed to persist much longer – up to several hours in exceptional cases. The longest continuous observation of a rainbow was made at the Chinese Culture University at Yangmingshan, on 30 November 2017. The rainbow persisted for an incredible 8 hours and 58 minutes.

Sony ILCE-7RM3 + FE 24–70 mm *f*/2.8 GM: 2 s; *f*/16; 24 mm; ISO-50.

The wrath of Storm Eleanor

Bill Brooks

Strong winds whip up large waves at Newhaven on the south coast of England. The World Meteorological Organisation employs a 10-point scale, running from 0 to 9, to describe the sea state – that is, the characteristics of the sea surface and the typical height of the waves. At zero on the scale, the sea is described as being glassy in appearance with a complete absence of waves. At point 5 on the scale the sea is described as being rough, having a typical wave height of 2.5 to 4.0 m (8¼ to 13 ft). At 9 on the scale, the sea state is described as phenomenal, with a wave height of over 14 m (46 ft). The very largest ocean waves occur when strong winds are able to act for a long period of time over a large distance, or 'fetch', of open water. In northwest Europe, the largest swells tend to be associated with vast low pressure systems tracking steadily east or northeast through the North Atlantic. On exposed west-facing coasts in countries such as Portugal and Ireland, wave heights may occasionally exceed 20 m (65½ ft), especially at certain localities where the coastal topography acts to amplify the waves as they approach the coastline.

Canon 5D Mark IV + EF 100–400 mm: 1/600 s; *f*/11; 400 mm; ISO-500.

Sunshine and shadows

Chris Brown

Sunshine, shallow fog and frost make for a morning of contrasts at Scotney Castle in Kent, England. Crepuscular rays are cast in the layer of fog by the shadow of trees. Strong sunshine has already melted the frost in places, such that the vibrant green of the sun-warmed grass contrasts strongly with icy coldness persisting in the shady frost hollows. True frost hollows are generally located in depressions in hilly areas, where cold air collects but cannot drain out easily. Sometimes these can be enhanced by human activity – for example, where the entrance to a valley is blocked by a railway embankment. The cold air, being denser than its surroundings, tends to sink into these hollows so temperatures tend to fall much lower than in surrounding areas on clear, calm nights. The resulting pools of cold air may even persist throughout the following day in the right conditions, especially if trees and surrounding hills help to shade the area from direct sunshine, leading to supressed daytime maximum temperatures. The cold air is dissipated when stronger winds arrive and cause mixing with the surrounding warmer air.

Hasselblad L1D-20c + 28 mm *f*/2.8: 1/160 s; *f*/4.5; 10.26 mm; ISO-100.

Solitary steel star

Yuriy Stolypin

The top of a skyscraper protrudes from a layer of fog in St Petersburg, Russia. Fog is a cloud of tiny liquid water droplets that is thick enough to reduce the visibility to less than 1 km (¾ mile). It is essentially a ground level stratus cloud. Fog contains up to 0.5 ml of water within each cubic metre of air. This means that if you concentrated the water from an area of fog filling an entire Olympic swimming pool, you would only be left with approximately 1.25 l (42 oz) of water. With over 200 days of fog each year, the Grand Banks off the coast of Newfoundland, Canada, is the foggiest place on Earth.

Hasselblad L1D-20c on DJI Mavic 2 pro drone: 1/400 s; *f*/5.6; 10.26 mm; ISO-100.

Dandelion dew stars

Chris Adams

Drops of dew on a dandelion seed head sparkle like jewels in the morning sunlight at Virginia Beach, Virginia, USA. Dew consists of very small droplets of water on surfaces such as grass and other plants, usually seen during the morning after nocturnal cooling of the air near the surface under clear and calm conditions. When the air is cooled to below the dew point temperature, some of the water vapour condenses out as dew. If the air is cold enough, this dew may subsequently freeze. Dew is actually a form of precipitation. Although the amount of water deposited on any given night is slight, collectively it may account for a substantial fraction of the total annual precipitation in some very arid regions.

Nikon Coolpix P900 + 83X 43–357 mm optical zoom lens.

Beauty before the storm

Mikhail Shcheglov

A double rainbow forms as a shower approaches the coastline at the southern tip of Iceland, near the village of Vik. A complete rainbow as seen from the air forms a full circle centred around the antisolar point, which is the point directly opposite the sun from the observer's perspective. However, near ground level, much of the bow is obscured from view because it is below the horizon. When the sun is near the horizon, as in this case, around half of the complete circle may be visible, with the centre of the bow rising high above the horizon. When the sun is at a greater angle above the horizon, the bow appears shallower because less of the complete circle is visible above the horizon. When the sun is higher than about 42° above the horizon, the rainbow is not visible since all of the bow would be below the horizon (the radius of the primary bow being approximately 42°).

Nikon D800 + AF-S NIKKOR 14–24 mm: *f*/2.8 G ED; 1/30 s; *f*/14; 14 mm; ISO-100.

Seeing double

Neil Partridge

A spectacular double rainbow appears as sunshine follows a heavy shower at Calanais on the Isle of Lewis, Scotland. Rainbows form due to the refraction and reflection of sunlight as it passes through raindrops. The secondary, outer bow, in which the colour banding is in the opposite direction to the primary bow, occurs when light is twice reflected off the back of the raindrops. The secondary bow always lies nine degrees outside of the primary. Between the two bows, especially near the horizon, a phenomenon known as 'Alexander's dark band' is apparent – a region in which the sky appears slightly darker than it does outside of the secondary bow and inside of the primary bow. This is because reflected light tends to brighten the sky outside of the secondary bow and inside of the primary bow. The phenomenon is named after Alexander of Aphrodisias who first described it in AD200.

Olympus E-M1 Mark II + Olynpus M.12–40 mm *f*/2.8: 1/400 s; *f*/10; 12 mm; ISO-200.

A sea of clouds

Gareth Mon Jones

Mountains in Snowdonia National Park, Wales, rise above an ethereal layer of early morning fog at the end of a cold night. 'After a long and sketchy hike up The Lliwedd on the flanks of Snowdon, trying to shoot the Milky Way, I finally managed to get above the temperature inversion that had hampered my shoot, as the night lost its battle against the rising sun. I was inspired to take a long exposure of the mesmerizing cloud as it flowed like liquid, leaving Moel Siabod just about peaking through.'

Nikon D810 + 24-70 mm *f*/2.8: 30 s; *f*/8; 70 mm; ISO-250.

A cap of clouds

Frolova Yulia

A cap of clouds crowns the summit of Vilyuchinsk volcano in Russia. Clouds often form preferentially over hills and mountains. This is because air is forced to rise as it flows over the higher ground. As the air rises it expands and cools. Cooler air can hold less water vapour than warmer air, and so with enough cooling, the air eventually reaches its dew point temperature – the temperature at which water begins to condense out as tiny water droplets – and a cloud is born. If the air is initially very moist, only a small amount of lifting will be needed, whereas dry air requires a lot of lifting to reach the dew point temperature and for a cloud to form. On some occasions, the opposite situation arises, and mountain tops are observed to protrude above a layer of cloud. This tends to happen when there is a layer of much drier air aloft, especially if moisture and clouds at lower levels are trapped below a temperature inversion (a layer in which the temperature increases with height in the atmosphere) such that the clouds spread out into a well-defined layer instead of rising up the mountain slopes. In the photograph shown here, a temperature inversion likely explains the extensive, lower layer of cloud above which the mountain top protrudes, whilst forced ascent of a slightly drier, overlying layer of air has resulted in the smaller cap of cloud near the mountain's summit.

Canon EOS 5D Mark III + EF 24–70 mm *f*/2.8L II USM: 1/3200 s; *f*/2.8; 50 mm; ISO-100.

Wind and wave power

Jay Birmingham

Huge waves crash into the sea wall at Porthcawl in South Wales. The Beaufort scale is an empirical measure for describing the wind speed that was originally based on the effects of the wind on the sails of ships, but subsequently changed so as to relate to the observed sea conditions. The scale runs from 0 (calm) to 12 (hurricane). At force 4, a moderate breeze, frequent white horses (breaking waves) are seen on the sea surface, whereas at force 9, a severe gale, high waves with dense streaks of foam are blown in the direction of the wind and spray reduces visibility. Land-based descriptors were added later, including 'whole trees in motion; inconvenience felt when walking against the wind' (force 7, near gale) and 'slight structural damage (chimney pots and slates removed)' (force 9, severe gale).

Canon EOS 6D Mark II + EF 70–200 mm *f*/2.8L IS II USM: 1/1000 s; *f*/13; 70 mm; ISO-500.

Wave pile-up

Thomas Rutherford

Huge waves crash against the sea wall at Seaham in County Durham, England, as gale-force north-easterly winds batter the coastline. Coastlines bordering the North Sea, including the east coast of the UK and north-facing coasts of the Netherlands and Belgium, are prone to occasional severe storm surges. These events are associated with a combination of spring high tides and northerly gales as a strong low pressure system originating from the North Atlantic Ocean travels east or southeast through the North Sea. The strong winds cause the waters to 'pile up' as they are driven south across the North Sea, an effect exacerbated by the increasing shallowness of the sea with southward extent. Furthermore, the low atmospheric pressure causes a rise in the sea level, which can be envisaged as a very shallow bulge in the sea surface underneath the area of low pressure, contributing further to the surge.

Canon EOS 5D Mark IV + EF 300 mm *f*/2.8L IS USM: 1/1600 s; *f*/11; 300 mm; ISO-100.

The frozen beach

Kieran Brimson

Snow coats the sand right down to the water line on Hayle beach in west Cornwall, England. Lying snow is comparatively rare along the immediate coastline of the UK, especially in the southwest, because the surrounding waters are relatively warm and so tend to keep coastal temperatures above freezing. However, this situation can occasionally change when extremely cold air originating from eastern Europe or western Russia is blown into the UK on easterly winds, usually in conjunction with a persistent area of high pressure centred over Scandinavia.

Nikon D800 + 10–24 mm *f*/3.5-4.5: 1/320 s; *f*/11; 10 mm; ISO-100.

Full house of rainbows

Carlos Castillejo

A spectacular display of multiple rainbows at Moskenesoy in the Lofoten Islands, Norway. In addition to the primary and secondary bows, a prominent primary reflection rainbow can be seen, caused by refraction of sunlight that has first been reflected in the surface of the fjord. Reflected bows can also be seen in the water of the lake. 'We were on holiday in the Lofoten Islands to enjoy the wonderful Norwegian landscapes and the northern lights. Not in my wildest dreams could I have imagined witnessing this spectacular display of rainbows. We were walking along the beach when at the bottom we could see how the sunlight filtered through the clouds and struck the fjord, which was mirror-like due to the lack of wind.'

Canon EOS 7D + Canon 17–55 mm: 1/60 s; *f*/5.6; 17 mm; ISO-250.

Bridge to nowhere

Artur Szczeszek

Clifton Suspension Bridge in Bristol, England, disappears into a thick fog. Fog is said to be dense when the visibility is reduced to below about 100 m (330 ft). When it becomes this thick it constitutes a significant hazard. The very lowest visibilities, however, tend to occur when fog is mixed with pollutants such as soot particles, creating 'smog'. When domestic and industrial coal use was widespread, smog was a relatively common hazard, especially in larger industrial cities. Today a different kind of smog, photochemical smog, is more common. This occurs when nitrogen oxides, hydrocarbons and volatile organic compounds in the atmosphere, released by motor vehicles and industry (and, in the case of the volatile organic compounds, by vegetation) react with sunlight, resulting in airborne particles and ground-level ozone. Los Angeles, USA, is particularly susceptible to photochemical smogs due to a combination of its location and climate. Mountains rise on three sides of the city and tend to shelter it, whilst a cool ocean on the fourth side acts to supress the vertical transport of air, allowing pollutants to accumulate to greater concentrations within a shallow layer near the ground. Abundant sunshine then provides ideal conditions for the photochemical reactions that create the smog to take place.

Nikon D7100: 6 s; *f*/19; 38 mm; ISO-100.

Spectres in the fog

Richard Fox

A fogbow and Brocken spectre appear as the sun breaks through banks of fog surrounding Meall Garbh, Perthshire, Scotland. Fogbows are an optical phenomenon arising from the diffraction of sunlight by small and uniformly-sized water droplets. When the observer is looking down from an elevated position their shadow may be cast on the fog or cloud bank below, appearing magnified – an effect known as the Brocken spectre. The delicately coloured rings surrounding the shadow of the photographer, visible to the left of centre in this image, are called a glory.

Sony ILCE-7RM3 + EF 16–35 mm *f*/4L IS USM: 1/125 s; *f*/11; 16 mm; ISO-100.

Diamond in the sky

Michal Krzysztofowicz

This complex display of halo phenomena is over the British Antarctic Survey's Halley Research Station in Antarctica. Visible are the 22° and 46° halos (rings forming at a radius of approximately 22° and 46° around the sun), parhelia (also known as 'mock suns' or 'sun dogs', visible as bright patches on the 22° halo at the same elevation as the sun), part of the parhelic circle (the faint, horizontal arc passing through the position of the sun), an upper tangent arc (the small arc in contact with the top of the 22° halo) and a sun pillar (the vertical column of light extending above and below the sun). In this case the halo display lasted most of the day due to the presence of diamond dust near ground level. The amount of diamond dust varied throughout the day; this photo was taken when the amount was at a maximum, and the halo display was consequently at its most spectacular.

Nikon D750 + Nikon AF-S Nikkor 14–24 mm *f*/2.8 G ED: 1/1250 s; *f*/8; 14 mm; ISO-100.

Tsunami of fog

Jaro Fagan

Fog shrouds the slopes of Mount Brandon on the Dingle Peninsula of southwest Ireland. In this photo, the effects of topography on the formation of clouds and fog are clearly apparent. Air is forced to rise on the windward side of the ridge, at the right-hand side of the image. As the air rises it expands and cools. Eventually, condensation of water vapour occurs, resulting in fog which shrouds the upper, windward parts of the ridge. This type of fog is sometimes known as 'upslope' fog. As the air crests the ridge and begins to descend again on the opposite side, it warms. As a result the fog begins to evaporate, leaving the leeward side of the ridge largely fog-free.

Canon EOS 6D + EF 24–70 mm *f*/4L IS USM: 3/5 s; *f*/14; 28 mm; ISO-160.

Cathedral in the marshes

Amanda Burgess

A misty, frosty morning over the marshes at Blythburgh, Suffolk, England. The church of Holy Trinity is known locally as the 'Cathedral of the Marshes', towering as it does above the flat landscape of the area. Here, it stands silhouetted against the dawn sky and surrounded by a shallow layer of radiation fog. This type of fog forms on calm, clear nights due to cooling of the land surface. The land then cools the adjacent air by conduction, to the extent that the lowest layers of air in closest contact with the ground become much cooler than the overlying layers. The resulting temperature inversion helps to trap the fog that forms within the cold air when it reaches its dew point temperature.

Sony ILCE-7M3 + FE 24–105 mm *f*/4 G OSS: 1/250 s; *f*/5.6; 24 mm; ISO-100.

Piercing through the clouds

Yuriy Stolypin

A skyscraper penetrates through the top of a layer of fog over St Petersburg, Russia. Advection fog occurs when moist air passes over a cool surface due to the action of the wind. The air closest to the surface is cooled to its dew point temperature, at which point condensation occurs. Advection fog is particularly common along coasts where the water temperatures are relatively low. It can be persistent, but it tends to 'burn off' when it moves over warmer surfaces, such as over land during the daytime. This means that coastal areas may be plagued by persistent fog, whilst locations only a few miles inland bask in warm sunshine. In the UK, advection fog forming over the North Sea is often called sea fret or haar. Advection fog can also form when moist air is cooled as it moves over snow-covered ground during the winter.

Hasselblad L1D-20c on DJI Mavic 2 pro drone: 1/400 s; *f*/5.6; 10.26 mm; ISO-100.

Mist in the morning

Vu Trung Huan

Areas of mist and shallow fog surround hillside tea plantations in the Tan Son District of Phu Tho Province, Vietnam. So what is the difference between fog and mist? Both are composed of suspended water droplets, but fog reduces the surface visibility to less than 1 km (¾ mile), whilst mist occurs when the visibility is reduced somewhat, with a relative humidity above 95%, but the visibility remains greater than 1 km (¾ mile). Usually, we observe mist when the visibility drops to 10 km (6 miles) or less; however there is no uniform definition around the world for the maximum visibility at which mist is still reported.

Canon EOS 5D Mark IV + EF 70–200 mm *f*/2.8L IS II USM: 1/30 s; *f*/11; 70 mm; ISO-50.

Supernumeries

Dani Agus Purnomo

A rainbow frames an isolated hilltop tree amongst tea fields. Towards the centre of the bow, a phenomenon known as supernumeraries is apparent – additional, faintly coloured bands on the inner edge of the main bow. These bands are dominated by green to purple colours, with this part of the colour sequence of the main bow sometimes being repeated several times. Supernumeraries tend to occur when the raindrops are relatively small and uniformly sized.

Nikon D3200 + TAMRON SP AF 150–600 mm *f*/5-6.3 VC USD A011N: 1/80 s; *f*/8; 280 mm; ISO-100.

Blue clouds before dawn

Owen Humphreys

A fine display of noctilucent clouds fills the pre-dawn sky above St Mary's lighthouse at Whitley Bay, England. Noctilucent clouds occur at heights far exceeding those of normal clouds – typically between 75 and 85 km (46½ and 53 miles) above ground level. They are too faint to be seen in the daytime and so are generally observed in the twilight sky before dawn or after sunset, and usually in the summer months between latitudes of about 45° and 80° north and south of the equator. Noctilucent clouds glow because their great altitude means that they remain sunlit even when the lower layers of the atmosphere and the Earth's surface are in shadow – the word noctilucent literally means 'night-shining'. These clouds typically have a silvery grey, bluish coloration and they frequently exhibit a complex structure, such as the rippling effects evident in this image.

Sony ILCE-9 + FE 24–70 mm *f*/2.8 GM: 30 s; *f*/9; 70 mm; ISO-200.

Typhoon makes landfall

Peng-Gang Fang

A ship runs aground as a typhoon makes landfall at Kaohsiung City. Typhoons (also known as hurricanes when they form in the Atlantic and northeast Pacific, and cyclones in the South Pacific and Indian Ocean) start life as clusters of thunderstorms. As the thunderstorms grow and merge together over warm ocean waters, moist air is drawn in towards the centre of the stormy region where the pressure begins to fall. Over time, these inflowing winds start to rotate as the pressure continues to fall near the centre of the storm system. The rotating, inflowing winds supply the storm with even more moisture and energy, causing it to strengthen further and the central pressure to decrease even more rapidly, further increasing the inflow of moist air. A positive feedback is thereby set up which, under the right conditions, allows the storm to become intense. When the wind speed near the centre reaches a sustained 73 mph (118 kph), a typhoon is born.

A kaleidoscope of colours

Rafal Kaniszewski

Dense cirrus clouds accompanied by islands of altocumulus make for a spectacular sunset over Rosendal in Norway. Cirrus clouds are composed of ice crystals and they are typically located at a height of between 6 and 12 km (3¾ and 7½ miles) above ground level in the middle latitudes. The name 'cirrus' means lock or tuft of hair in Latin. The characteristic fibrous, trailing parts of cirrus clouds are caused by ice crystals slowly falling from the main cloud mass. These falling ice crystals are often drawn out into long streaks by strong winds aloft. The falling ice crystals sublimate – they revert directly to water vapour – long before reaching the ground, and so the cirrus streaks tend to taper off below the cloud base, as can be seen near the top, centre of the photograph. These tapering streaks are called 'virga'.

Nikon D850 + 14–24 mm: 1/3 s; *f*/11; 14 mm; ISO-100.

The Helm Wind

Stephan Brzozowski

An ominous roll of low cloud hovers over the Eden Valley in Cumbria, England. The Helm is England's only named wind, occurring when strong northeasterly winds blow across the high fells of the northern Pennines to the northeast of the Eden Valley. The Helm Bar occurs when low cloud evaporates as air descends on the leeward side of the hills, leaving a prominent cap of clouds shrouding the hilltops. The downslope winds associated with the Helm may be fierce but they are also very localized. Sometimes, as here, additional rolls of cloud occur over the Eden Valley, due to parallel bands of ascending and descending air extending some way to the lee of the main ridge of hills. Another phenomenon visible in this image, within the tenuous cloud near the edge of the main cloud roll, is a corona. Coronas are coloured rings of light around the sun which form due to the diffraction of sunlight by tiny water droplets within the cloud.

Nikon D810 + 18–35 mm *f*/3.5–4.5: 1/1600 s; *f*/8; 20 mm; ISO-200.

Ring of frost

Ales Krivec

A prominent 22° halo around the sun, which in this case has formed due to the presence of tiny ice crystals known as diamond dust, is suspended in the air near ground level in the mountains of Slovenia. The 22° halo is so-called because it forms a ring with a radius of about 22° around the sun or moon. Diamond dust forms when temperatures are well below freezing (usually less than -10°C (14°F). It is associated with a process known as deposition, where water in its gaseous state (water vapour) converts directly to a solid (the ice crystal) without going through the liquid phase. Individual crystals can sometimes be discerned when they sparkle in the sunlight, as can be seen here near the upper parts of the halo.

Nikon D800 + 14–24 mm *f*/2.8: 1/80 s; *f*/16; 19 mm; ISO-100.

A waterfall of fog

James Loveridge

Photographed over the Jurassic Coast of Dorset, England, localized differences in the density of air are made visible by the presence of a shallow layer of fog. The fog has formed over the land due to radiative cooling of the land surface under clear skies. The cold land surface in turn cools the lowest layers of air that are in contact with it by conduction, forming a shallow layer of fog where the air is cooled to its dew point temperature. The cold, foggy air is denser than the surrounding air, and so it tends to flow towards topographical low points. In this case, when the cold, foggy air reaches the coast, it spills over the cliffs in a spectacular 'waterfall', before dissipating near the foot of the cliffs, likely due to mixing with the warmer air overlying the waters of the English Channel.

DJI FC 300 X: 1/1000 s; *f*/2.8; 20 mm; ISO-100.

Fog in the valley

David Hendry

There are many different types of fog, one of which is valley fog, here on the River Clyde in Glasgow, Scotland. This is actually a type of radiation fog – fog that forms due to cooling of the land surface and overlying air on calm, clear nights. However, as the name suggests, it forms within valleys where cold, dense air collects under the right conditions (light winds, clear sky and sufficient moisture in the air). Valley fog can last for several days at a time, particularly during the winter, when it may hold temperatures below freezing point even by day.

Sony A65 + Sony 30 mm *f*/2.8 DT: 1/640 s; *f*/8; 30 mm; ISO-100.

Floating in the fog

Chirag Khatri

Skyscrapers rise above a layer of fog over Dubai, United Arab Emirates. Although fog is rather uncommon in Dubai, occurring only a few days a year, when it does form the results can be spectacular. Most occur during the autumn and winter, when longer nights permit greater cooling of the warm and relatively moist air masses that often reside across the region. Dust and airborne pollutants may encourage fog formation by providing a plentiful supply of condensation nuclei on which the water droplets comprising fog can grow, once the air has cooled to its dew point temperature.

Nikon D7200 + Tamron SP 15–30 mm *f*/2.8 Di VC USD (Nikon F): 1/10 s; *f*/8; 15 mm; ISO-100.

Celestial rays

Alan Haynes

Dark clouds and crepuscular rays combine to dramatic effect above the Black Church at Búðir, Iceland. Crepuscular rays commonly occur when sunlight passes through gaps in the cloud, forming parallel bands of light and shade. These are rendered visible by scattering of sunlight by particles suspended in the atmosphere, such as small water droplets or dust. Although parallel, the rays appear to diverge due to the effect of perspective – the rays appear to converge towards a single point with increasing distance from the observer. Many different types of particle can be suspended in the atmosphere and these particles are known collectively as aerosols. The amount and type of aerosol depends largely on the source of the air. For example, air originating over oceans often contains large amounts of sea salt aerosol, derived from the bursting of bubbles in sea foam, whereas air originating over arid continental areas is often rich in dust aerosol, because strong winds frequently lift particles from the dry, sparsely vegetated land surfaces in these areas. Other sources of aerosol are volcanic eruptions and human activity such as biomass burning.

Nikon D800 + Nikon AF-S NIKKOR 24–120 mm *f*/4 G ED VR: 1/800 s; *f*/8; 86mm; ISO-400.

Colours of the aurora

Graeme Whipps

The aurora borealis, also known as the northern lights, takes its name from the Latin for dawn. The aurora occurs when charged particles emitted by the sun are deflected by the Earth's permanent magnetic field and interact with the upper atmosphere. They light up the night sky with slowly evolving rays of beautiful colour. Although most commonly observed at high latitudes, when a large geomagnetic storm occurs the phenomenon may be seen at lower latitudes, as in this example over Aberdeenshire, Scotland. The largest geomagnetic storm recorded to date, known as the 'Carrington Event', occurred on night of 1st September 1859. In this case the aurora borealis was reportedly visible as far south as the Caribbean. The different colours in the aurora are associated with interaction of the charged particles with different gases in the upper atmosphere. For example, green and red colours are characteristic of oxygen, with the colour depending on height, whilst blue or purple hints are associated with nitrogen.

Canon 6D + Sigma 24 mm *f*/1.8 EX DG: 10 s; *f*/2.8; 24 mm; ISO-1600; Panorama of two images stitched using Hugin.

A golden layer of fog

Michael Epel

A layer of fog shrouds the Val d'Orcia in Tuscany, Italy, at sunrise after a clear and cold night. Radiation fog forms due to nocturnal cooling of the ground under clear skies. The ground cools the air immediately above it by conduction, so that a relatively shallow layer of cold air forms just above the ground. When the air within this layer cools to its dew point temperature, fog forms. Light winds are needed for radiation fog, because strong winds would otherwise mix the shallow layer of cold air with warmer, drier overlying layers of air, so that temperatures would be prevented from falling very far. However, a very small amount of wind can actually be helpful for the formation of fog – when there is absolutely no wind at all, water vapour is often deposited instead as a heavy dew.

Phase One + Schneider Kreuznach 80 mm; 1/20 s; f/16; 80 mm; ISO-35.

Above the inversion

Richard Fox

Under normal circumstances, temperature decreases with height within the troposphere (the lowest layer of the atmosphere within which the majority of our weather occurs). However, this situation can be reversed under clear, calm conditions at night because air in contact with the ground cools more rapidly than the overlying air. A layer in which the temperature increases with height is known as a temperature inversion. Sometimes, the effects of a temperature inversion are rendered visible by the formation of fog within the cooled air mass below the inversion, as in this example over Dartmoor in Devon, England. On some occasions the fog may persist all day over low ground, suppressing temperatures there, whilst any hilltops poking out of the fog enjoy, perhaps counterintuitively, much milder temperatures.

Sony ILCE-7RM3 + EF 16–35 mm *f*/4L IS USM: 1/125 s; *f*/11; 35 mm; ISO-100.

Tidal waves of fog

Aleksandra Karptsova

Waves of a different kind – fog – shroud the harbour and surrounding coastline at Vladivostok, Russia. Coastal regions with relatively cool waters are frequently subject to fog as the overlying air is cooled to its dew point temperature on contact with the water. A well-known example of this phenomenon is the sea fog that frequently affects San Francisco and other locations along the Californian coastline in the USA. These fogs are particularly common in summer when the dew point temperature of the air tends to be higher than the sea surface temperature, so that a large bank of fog forms over the coastal waters. As the land heats up during the day the pressure falls, causing an onshore breeze that pushes the foggy marine air inland. In San Francisco itself, an average of five hours of fog is experienced every single day in the summer, though there are large local variations depending on proximity to the coastline and the subtleties of the local coastal topography.

Canon EOS RP: 1/50 s; *f*/3.2; 100 mm; ISO-3200.

A river of clouds

Sabrina Garofoli

A spectacular 'river of clouds', following the Adda river in Lombardy, Italy is further enhanced by an array of beautiful crepuscular rays, cast by the shadows of trees along the banks of the river and in the surrounding fields. On clear and calm nights, especially during autumn and winter, radiation fog may form. Overnight, the land surface loses heat by radiation. The air in contact with the ground is then cooled by conduction. If the air is cooled sufficiently (to the dew point temperature), condensation of water vapour occurs, resulting in fog. The cool, dense air will always sink to the lowest point it can, which is why it is common to see fog in river valleys and other low-lying areas. However, sometimes it is possible to witness a 'river of clouds' closely following an actual river, usually when the river water is relatively warm so that it provides an additional source of moisture for fog formation.

Canon EOS 6D + EF 70–200 mm *f*/4L USM: 1/200 s; *f*/11; 70 mm; ISO-160.

Lunar halo

Mikhail Kapychka

A complex lunar halo display forms over Mogilev, Belarus. Visible are the 22° halo, part of the paraselenic circle (the faint, horizontal arc passing through the position of the moon) and a particularly prominent upper tangent arc (the bright arc in contact with the top of the 22° halo), which displays noticeable coloration. Halos form due to the refraction of sunlight or moonlight by hexagonal ice crystals orientated at many different angles. In this case, the ice crystals occur within a sheet of cirrostratus – a thin, but horizontally extensive layer of cloud at high altitude, which gives the sky a milky veil through which the stars can still be seen. The coloration of the upper tangent arc and halo, with red towards the lower and inner edges, respectively, arises because light of different wavelengths undergoes different amounts of refraction, with red light being refracted slightly less than blue light.

Canon EOS 5D Mark II + EF 16–35 mm *f*/2.8L USM: 30 s; *f*/4.5; 16 mm; ISO-800.

Sunlight reflections

Lisa Wood

Sunlight is made up of a spectrum of colours ranging from red to violet, with each having a slightly different wavelength. Air molecules scatter sunlight as it travels through the atmosphere, in particular the shorter wavelengths – the blue and violet light – which is why the sky appears blue to us. When the sun is lower in the sky, such as at sunset, the light must pass through more of the atmosphere and much more scattering occurs, such that much of the blue light is deflected away in different directions before reaching us. This means that comparatively more of the longer wavelengths – the red and orange light – reaches us at the surface. In order for the sky to appear vividly red or orange, however, medium and high-level clouds, such as the altocumulus lenticularis clouds seen here in Nairn, Scotland, are usually required, which further scatter and reflect the red light.

Panasonic Lumix DMC-TZ30: *f*/4.4; 51 mm, ISO-100.

Clouds in the stratosphere

Alan Tough

Beautifully coloured nacreous clouds glow in the evening twilight above Alloa in Scotland. In late January and early February 2016, unusually cold air in the stratosphere spread south from the polar regions towards the UK. This triggered sightings of rare and beautiful nacreous clouds, also known as polar stratospheric clouds. As their name suggests, these clouds form within the stratosphere – the layer of atmosphere that extends from approximately 10 to 50 km (6 to 31 miles) above ground level – and so they are located at much higher altitudes than the tropospheric clouds that produce most of our weather. Polar stratospheric clouds are composed mainly of ice and require extreme cold – temperatures around -85°C (-121°F) – in order to form. Because they are located at such high altitudes, they remain sunlit well after the sun has set near ground level, such that they may shine brilliantly in the twilight sky.

Canon PowerShot G9 + 7.4–44.4 mm: 1/2000 s; *f*/8; 37mm; ISO-400.

Actions to tackle climate change

JIM WATSON

The greenhouse gas emissions that cause climate change have continued to increase rapidly during the past few decades, largely driven by burning fossil fuels (coal, oil and gas), which make up 80% of the world's energy use. Whilst there is some optimism that global emissions may level off, it is too early to tell whether this is the case. Emissions in 2019 reached a record high of around 52 gigatonnes, if emissions from direct human-induced land use, land use change and forestry activities are excluded. During the past two decades, the drivers of increasing emissions have changed. Many industrialized countries have either stabilised or reduced their emissions. For example, the UK's emissions have reduced by around 45% since 1990 by phasing out coal in the power sector, energy efficiency and industrial restructuring. Global emissions growth has been driven by the rapid emergence of middle-income countries – particularly China, whose emissions have increased by over 250% since 2000.

The key international agreement on climate action came into effect over 25 years ago. However, the focus on sharing out emissions reductions among industrialized countries did not deliver the significant reductions required. The Paris Agreement of 2015 took a different approach, based on voluntary, national plans submitted to the UN by both developed and developing countries. It included a pledge for climate neutrality by the second half of the 21st century, regular reviews of emission reduction plans, and a process of increased ambitions over time.

What has this delivered so far? The UN Environment Programme's (UNEP) 2020 assessment concludes that progress has been made. However, current pledges mean that average temperatures will still increase by around 3°C (5.4°F). This is much higher than the 'well below 2 degrees' goal called for in the Paris Agreement to avoid dangerous anthropogenic climate change. An increasing number of countries have targets for reducing emissions to 'net-zero', to help meet the Paris Agreement goal. They include legally binding targets for 2050 in the UK and France, for example. Net-zero means reducing emissions as much as possible, and then balancing remaining emissions with measures to remove greenhouse gases from the atmosphere. This includes preserving and expanding forests and farming practices that lock away carbon in the soil. If these net-zero plans are taken into account, and backed up by shorter term actions, UNEP estimates the average temperature increase will be around 2.5 °C (4.5°F). This is better – but still not enough.

What needs to be done to reduce our emissions? All sectors need to change. This includes the way electricity is produced, how we travel, the food we eat, the energy used by industry and the way we heat and cool our buildings. The most rapid changes so far have occurred in electricity generation. Many countries have provided financial support for renewable electricity, resulting in a steep increase in investment. This has driven down costs, especially of technologies that harness the sun and wind: solar and wind power. These renewable technologies, together with nuclear power and hydro power, generate around a third of the world's electricity. This share is set to rise. According to the International Energy Agency, around 90% of global power plant investment went into renewables in 2020.

The way we travel is also starting to change. In principle, electric vehicles have much lower greenhouse gas emissions than fossil fuel vehicles – as long as they are charged with a low carbon source of power. The market for electric vehicles is expanding rapidly. Sales of electric vehicles grew 43% in 2020 when compared to the previous year and this is being driven by ambitious targets in some countries. Shifting to low carbon trucks, ships and planes will be much harder, however. In the case of aircraft, there have been some experiments with small electric planes that can travel short distances. Yet air travel is likely to cause significant emissions for decades to come. These emissions, which have a greater impact on the climate because of complex interactions with the atmosphere, will need to be offset by removals.

It will also be more difficult to reduce emissions from some industries, especially those that use a lot of energy such as steel, cement and chemicals. There is hope that a combination of low carbon energy sources, more efficiency and shifts to lower carbon materials (e.g. wood for construction) will make significant progress. Hydrogen is being discussed as an option to reduce emissions from industry. The problem is that it is expensive to produce, and some methods of production still use fossil fuels. Whilst there are technologies that can capture and store most of the emissions from hydrogen production using fossil fuels, few plants have been built so far. Reducing emissions from buildings can be achieved through a range of technologies and measures – many of which are already in widespread use. They include better insulation and the use of low carbon electric heat pumps or district heating, where heat generated at a central location is distributed to multiple premises. Hydrogen could also be used for heating and cooling in countries with established gas pipeline networks, but implementation is at a very early stage. Climate change action also means changes to farming and land management. Changes in diet that involve less consumption of meat and dairy would have two benefits. First, it would reduce direct emissions from agriculture. Second, it would free up agricultural land so it can be repurposed for carbon removal, for example by planting trees.

The economic argument for acting sooner rather than later is well established. As the influential Stern Review argued in 2006, the costs of reducing emissions are very likely to be much lower than the costs of inaction. Action to reduce emissions can support new industries and create jobs – for example in the design and manufacture of low carbon technologies or in upgrading buildings to make them more efficient. The Covid-19 pandemic has had severe impacts on many economies. This creates an opportunity for governments to integrate climate action into their plans to help their economies recover. There are health benefits too. For example, a shift to low carbon vehicles not only reduces greenhouse gas emissions, it also reduces local pollution which can cause or exacerbate respiratory illnesses. Making homes more energy efficient can have health benefits, especially for poorer households that can't afford sufficient heating.

In many developing countries, health is a key driver of efforts to improve access to energy. In 2018, 789 million people did not have access to electricity, whilst 2.8 billion people did not have access to clean and efficient cooking appliances and fuels. Whilst climate change action will not necessarily close this access gap, low carbon technologies can play a significant role in doing so.

Even if the world achieves the Paris Agreement goals, adaptation to climate change will be required. It will vary widely by location and include changes to agricultural practices (e.g. through the use of more drought resistant crops), changes to infrastructure (e.g. to take into account sea level rise or changes in flooding) and early warning systems for extreme weather events. A striking feature of many climate action plans submitted to the UN by low income countries is that they often focus more on adaptation than on measures to reduce emissions. This is because their emissions tend to be very low, and because of their higher vulnerability to climate change impacts.

Whilst the world is not doing enough to tackle climate change, there are reasons for optimism. The costs of some of the key technologies have fallen rapidly, and that means their use is expanding quickly. Global climate politics are entering a more promising period, and many businesses are aiming to reduce their carbon footprints to zero. Citizens are also much more conscious of the choices they can make to reduce emissions.

To translate these signs of hope into real progress will require at least three conditions to be met. First, countries that have made ambitious long-term pledges need to back these up with short term actions to reduce emissions at the rate required. Second, more countries need to be persuaded to join them – including emerging economies that are likely to need financial and other support. Third, there is a need to address the losers as well as the winners with a political strategy for those industries and communities that are dependent on fossil fuels - and to ensure that costs and benefits are distributed fairly.

Bibliography

The following literature was referred to in the content development of the essays.

How and why Earth's climate is changing

Allan, R.P. and Soden, B.J., Atmospheric warming and the amplification of precipitation extremes. *Science*, 2008, 321 (5895): 1481–1484.

Cvijanovic, I., Lukovic, J. and Begg, J.D., One hundred years of Milankoviç cycles. *Nat Geosci*, 2020, 13: 524–525.

Harries, J., Brindley, H. and Sagoo, P. et al., Increases in greenhouse forcing inferred from the outgoing longwave radiation spectra of the Earth in 1970 and 1997. *Nature*, 2001, 410: 355–357.

Jackson, R., Notes and records: Eunice Foote, John Tyndall and a question of priority. *Roy Soc*, 2020, 74 (1): 74105–74118.

Le Treut, H., Somerville, R. and Cubasch, U. et al., Solomon, S., Qin, D. and Manning, M. et al. (eds.). IPCC 2007: Historical Overview of Climate Change. *Climate Change 2007: The Physical Science Basis.* Contribution of WGI to Fourth Assess Rep IPCC.

Stocker, T.F., Qin, D. and Plattner, G.-K. et al. (eds.), 2013. *IPCC 2013: Climate Change: The Physical Science Basis*. Contribution of WGI to Fifth Assess Rep IPCC.

Tierney, J.E., Poulsen, C.J. and Montañez, I.P. et al., Past climates inform our future. *Science*, 2020, 370 (6517).

Tierney, J.E., Zhu, J. and King, J. et al., Glacial cooling and climate sensitivity revisited. *Nature*, 2020, 584: 569–573.

Our climate and extreme weather

Abram, N.J., Henley, B.J., Sen Gupta, A. et al., Connections of climate change and variability to large and extreme forest fires in southeast Australia. *Commun Earth Environ*, 2021, 2 (8).

Bhatia, K.T., Vecchi, G.A., Knutson, T.R. et al., Recent increases in tropical cyclone intensification rates. *Nat Commun*, 2019, 10 (1): 1–9.

Ciavarella, A., Cotterill, D. and Stott, P. et al., Siberian heatwave of 2020 almost impossible without climate change, WWA, 2020: www.worldweatherattribution.org/siberian-heatwave-of-2020-almost-impossible-without-climate-change.

Dunn, R.J.H., Stanitski, D.M., Gobron, N. et al. (eds.), Global Climate [in 'State of the Climate in 2019']. *Bull Am Meteorol Soc*, 2020, 101 (8): S9–S127.

Goss, M., Swain, D.L., Abatzoglou, J.T. et al., Climate change is increasing the likelihood of extreme autumn wildfire conditions across California. *Environ Res Lett*, 2020, 15 (9): 094016.

Knutson, T., Kossin, J.P. and Mears, C. et al., Wuebbles, D.J., Fahey, D.W. and Hibbard, K.A. et al. (eds.), 2017. Detection and attribution of climate change. In: *Climate Science Special Rep: Fourth National Climate Assessment*, Vol I. US Global Change Research Program, Washington, USA.

Kossin, J.P., Knapp, K.R. and Olander, T.L. et al., Global increase in major tropical cyclone exceedance probability over the past four decades. *PNAS*, 2020, 117 (22): 11975–11980.

Lelieveld, J., Evans, J., Fnais, M. et al., The contribution of outdoor air pollution sources to premature mortality on a global scale. *Nature*, 2015, 525: 367–371.

van Oldenborgh, G.J., van der Wiel, K. and Sebastian, A. et al., Attribution of extreme rainfall from Hurricane Harvey. *Environ Res Lett*, 2017, 12 (12): 124009.

Vautard, R., van Aalst, M. and Boucher, O. et al., Human contribution to the record-breaking June and July 2019 heatwaves in Western Europe. *Environ Res Lett*, 2020, 15 (9): 094077.

Our climate and the natural world

Bai, X., van der Leeuwb, S. O'Brien, K. et al., Plausible and desirable futures in the Anthropocene: A new research agenda. *Glob Environ Change*, 2016, 39: 351–362.

Bar-On, Y., Phillips, R. and Milo, R., The biomass distribution on Earth. *PNAS*, 2018, 115 (25): 6506–06511.

Chen, I.C., Hil, J.K. and Ohlemueller, R. et al., Rapid range shifts of species associated with high levels of climate warming. *Science*, 2011, 333: 1024–1026.

Crutzen, P., The geology of mankind. *Nature*, 2002, 415: 23.

Duchenne, F., Thébault, E. and Michez, D. et al., Phenological shifts alter the seasonal structure of pollinator assemblages in Europe. *Nat Ecol Evol*, 2020, 4: 115–121.

www.frontiersin.org/articles/10.3389/fmars.2017.00158/full; www.ipbes.net/global-assessment; www.oxfam.org/en/press-releases/forced-from-home-eng; www.researchgate.net/publication/5609980_Death_toll_exceeded_70000_in_Europe_during_the_summer_of_2003; www.royalsocietypublishing.org/doi/pdf/10.1098/rspb.2020.1432; www.theverge.com/2019/9/25/20882502/marine-heatwaves-oceans-united-nations-ipcc-report-climate-change; WWF-Australia, *Australia's 2019-2020 bushfires: the wildlife toll.* Interim Rep, 2020: www.wwf.org.au/news/news/2020/3-billion-animals-impacted-by-australia-bushfire-crisis#gs.dnhsom

Our climate and Earth's frozen water

AMAP, 2017, *Snow, Water, Ice and Permafrost in the Arctic* (SWIPA) 2017, AMAP, 2017, pp.xiv + 269.

King, M.D., Howat, I.M. and Candela, S.G. et al., Dynamic ice loss from the Greenland Ice Sheet driven by sustained glacier retreat. *Commun Earth Environ*, 2020, 1 (1).

National Snow and Ice Data Center (NSIDC)

Sasgen, I., Wouters, B. and Gardner, A.S. et al., Return to rapid ice loss in Greenland and record loss in 2019 detected by the GRACE-FO satellites. *Commun Earth Environ*, 2020,1(8).

Slater, T., Hogg, A.E. and Mottram, R., Ice-sheet losses track high-end sea-level rise projections. *Nat Clim Chang*, 2020, 10: 879–881.

Slater, T., Lawrence, I.R. and Otosaka, I.N. et al., Review Article: Earth's ice imbalance. *The Cryosphere*, 2021,15: 233–246.

Žebre, M., Colucci, R.R. and Giorgi, F. et al., 200 years of equilibrium-line altitude variability across the European Alps (1901–2100). *Clim Dyn*, 2021, 56: 1183–1202.

Our climate and the ocean

Masson-Delmotte, V., Zhai, P. and Pörtner, H.O. et al. (eds.). IPCC, 2018: Summary for Policymakers. In: *Global warming of 1.5°C.* An IPCC Special Rep on the impacts of global warming of 1.5°C above pre-industrial levels and related global greenhouse gas emission pathways, in the context of strengthening the global response to the threat of climate change, sustainable development, and efforts to eradicate poverty. WMO, Geneva, Switzerland, 32pp.

Gruber, Nicolas, D. Clement, B.R. and Carter, R.A. et al. The oceanic sink for anthropogenic CO2 from 1994–2007. *Science*, 2019, 363: (6432): 1193–1199.

Pörtner, H.-O., Roberts, D.C. and Masson-Delmotte, V. (eds.), 2019. IPCC, 2019: Summary for Policymakers. In: *IPCC Special Rep on the Ocean and Cryosphere in a Changing Climate*.

Stocker, T.F., Qin, D. and Plattner, G.-K. et al. (eds.). *IPCC 2013: Climate Change 2013: The Physical Science Basis.* Contribution of WGI to the Fifth Assess Rep of the IPCC.

Actions to tackle climate change

BEIS (2021) Final UK greenhouse gas emissions national statistics: 1990–2019; www.gov.uk/government/statistics/final-uk-greenhouse-gas-emissions-national-statistics-1990-to-2019

International Energy Agency data for 2018: www.iea.org/data-and-statistics

International Energy Agency (2020) Renewables 2020. Paris: OECD/IEA. www.iea.org/reports/renewables-2020

Global sales of electric cars accelerate fast in 2020 despite pandemic. *The Guardian*,19 Jan 2021: www.theguardian.com/environment/2021/jan/19/global-sales-of-electric-cars-accelerate-fast-in-2020-despite-covid-pandemic

Lee, DS et al., 2021. The contribution of global aviation to anthropogenic climate forcing for 2000 to 2018. *Atmos Environ* 244:117834.

Sustainable Energy for All, 2020. SEforAll Analysis of SDG7 Progress – 2020; https://www.seforall.org/data-stories/seforall-analysis-of-sdg7-progress-2020

Vivid Economics and Imperial College, 2017. International comparisons of heating, cooling and heat decarbonisation policies.www.gov.uk/government/publications/international-comparisons-of-heating-cooling-and-heat-decarbonisation-policies

Watts, N. et al., 2021. The 2020 report of the Lancet Countdown on health and climate change: responding to converging crises. *The Lancet*. 397 (10269): 129–170.

United Nations Environment Programme (2020) Emissions Gap Rep 2020: www.unenvironment.org/emissions-gap-report-2020

Glossary

Numbers refer to page numbers for essays and image captions.

altocumulus – a medium level cloud, usually white or grey in colour, which tends to appear as bands or patches of wavy, rounded masses or rolls. 79, 108, 117, 127, 164, 186

anticyclone – an area of relatively high atmospheric pressure, characterized by outward-spiralling winds. In the northern (southern) hemisphere winds rotate around an anticyclone in a clockwise (anticlockwise) direction.

anvil – the flat, icy upper portion of a cumulonimbus cloud, so named because of its likeness to a blacksmith's anvil. 24, 52, 60, 62, 65

arcus cloud – a low level, horizontal cloud formation of which there are two types: shelf cloud and roll cloud. 17, 42

aurora borealis – a natural light display in the northern hemisphere, caused by charged solar particles being deflected by Earth's permanent magnetic field and interacting with gases in the upper atmosphere. 91, 124

Brocken spectre – the name given to the magnified shadow of an observer cast on to cloud or mist below. 152

cirrus – a thin and wispy high level cloud composed of ice crystals. 164

climate – the average weather conditions in a given place over a period of time, ranging from months to thousands or millions of years. The classical period for averaging variables (such as surface temperature, precipitation and wind) is 30 years, as defined by the World Meteorological Organization. 4, 128

climate change – a change in the state of the climate that can be identified by changes in the mean and/or the variability of its properties, and that persists for an extended period, typically decades or longer. Such changes can be natural or human-induced. 4, 6–7, 10–11, 48, 70–1, 88–9, 128–9, 188–9

cloud-to-ground lightning – an electrical discharge between a cloud and the Earth's surface. 14, 20, 46, 57, 66

crepuscular rays – streaks or beams of light radiating from the Sun, rendered visible by the scattering of sunlight by particles suspended in the atmosphere, such as small water droplets or dust. 69, 134, 175, 183

cumulus – a 'heaped-up' cloud formed from convection, often described as 'puffy', 'cotton-like' or 'cauliflower-shaped'. 95

cumulonimbus – a form of cumulus cloud with extensive vertical development, often producing heavy precipitation, thunder and lightning. 26, 41, 52

cyclone – an area of relatively low atmospheric pressure, characterised by inward-spiralling wind. In the northern (southern) hemisphere winds rotate around a cyclone in an anticlockwise (clockwise) direction. Can be mid-latitude (depressions) or tropical (includes hurricanes and typhoons). 11, 74, 128, 163

dust devil – a small, short-lived whirlwind made visible by sand, dust and debris picked up from the ground. Unlike tornadoes, dust devils originate from the surface and are not associated with thunderstorms. 32

fogbow – an optical phenomenon similar to a rainbow but with a largely white bow which appears in fog rather than rain. It arises from the interaction of sunlight with small and uniformly-sized water droplets. 152

freezing level – the height in the atmosphere at which air temperatures fall below freezing. Within the troposphere, temperatures typically decrease with increasing height above the surface. 27

global warming – the increase in global average temperature, linked to the anthropogenic increase in greenhouse gas concentrations in the atmosphere. 88, 129

graupel – a type of precipitation that forms when supercooled water droplets freeze on to a falling snowflake. Also called 'soft hail'. 27, 44, 56

greenhouse gas – any atmospheric gas that absorbs long-wave infrared radiation, thereby trapping heat in the lower atmosphere. Examples include carbon dioxide, methane and water vapour. 6–7, 10, 49, 71, 88, 89, 128, 188–9

haboob – a term used to describe any wind-driven sandstorm or dust storm in arid or semi-arid regions around the world. Haboobs are typically associated with the gusty outflow winds of a thunderstorm. 12, 30

hoar frost – a type of frost which forms when water vapour in the air freezes directly onto a surface to form ice crystals with a distinct feathery structure. 101, 104

lenticular cloud – a lens- or saucer-shaped cloud that forms as a result of airflow over a hill or mountain in stable conditions. 108, 117, 127, 186

mammatus cloud – a series of bulges or pouches hanging from the base of a cloud, typically a cumulonimbus, which form from sinking pockets of cold, moist air. 24, 41

monsoon – the seasonal shift in wind direction that brings alternate very wet and very dry seasons to much of the tropics and subtropics. 48, 74, 84

noctilucent cloud – a bluish or silvery coloured cloud consisting of ice crystals that occurs in the mesosphere and is only visible during the night. Not to be confused with polar stratospheric clouds or nacreous clouds. 162

parhelia – an optical phenomenon that consists of two bright patches either side of the Sun on the 22° halo. Also known as 'sun dogs', they form from the refraction of light through ice crystals. Sometimes only one parhelion (or 'sun dog') is visible. 153

rime ice – an opaque deposit of ice that forms when supercooled water droplets in the air freeze on contact with an exposed surface that has a temperature below freezing point. 101, 104, 110, 120, 123

sheet lightning – a term used to describe clouds illuminated by a lightning discharge where the actual lightning channel is either inside the clouds or below the horizon (i.e. not visible to the observer). 14

shelf cloud – a wedge-shaped arcus cloud that occurs along the leading edge of the rain-cooled air flowing out of a mature thunderstorm. 17, 30, 42, 54

stratocumulus – a low level cloud, usually white or grey in colour, which tends to appear as either patches or a sheet of rounded elements. 95

stratosphere – the highly stratified region of the atmosphere above the troposphere extending from about 10 km (ranging from around 6–8 km near the poles to around 16–18 km in the tropics) to about 50 km altitude. 10, 187

superbolt – a rare positive cloud-to-ground lightning bolt, which may discharge up to 1,000 times more energy than the average lightning strike. 57

supercell – a powerful thunderstorm with a deep rotating updraft, capable of producing tornadoes, large hail, damaging wind gusts (including downbursts) and torrential rain. 23, 44, 45, 51, 52, 58, 60, 61

tornado – a violently rotating column of air that extends between the base of a storm cloud and the ground, and which is usually made visible by the presence of a funnel-shaped cloud (a 'tuba') extending from the main cloud base of the storm. 23, 32, 34, 41, 45, 51

tropical cyclone/hurricane/typhoon - a tropical revolving storm with sustained wind speeds of more than 118 km/h (74 mph). It is called a hurricane in the North Atlantic and eastern North Pacific, a typhoon in the western North Pacific, and a tropical cyclone in the South Pacific and Indian Ocean. 11, 163

tropopause – the boundary between the troposphere and stratosphere. 10, 65

troposphere – The lowest part of the atmosphere, from the surface to about 10 km in altitude at mid-latitudes (ranging from around 6–8 km near the poles to around 16–18 km in the tropics), where clouds and weather phenomena occur. 52, 65, 179

weather – the state of the atmosphere at a specific time and place, with regards to variables such as wind strength and direction, temperature, precipitation, and pressure. 4, 10–11

Photographers' index

Photographer, page reference, titles and locations of images and photographer contact details.

Chris Adams p.137 Dandelion dew stars, Virginia Beach, Virginia, USA www.facebook.com/Cadams9705

Iain Afshar pp.126–127 Flying saucers, Chamonix, France www.thephotographyproject.co.uk @thephotographyprojectbristol

Simon Anderson pp.28–29 Painting in the sky, Eastbourne, England @_simonanderson_

Lauren Bailey pp.66–67 Early morning strike, El Paso, Texas, USA @lorigraceaz

Andrei Baskevich pp.124–125 Aurora over the forest, Paanayarvi National Park, Russia www.andreibaskevich.com @a.baskevich

Brian Bayliss p.122 Snow roller Marlborough, England @crownforestry

Hugo Begg pp.14–15 Lighting up the night, Trial Bay, Australia

Jay Birmingham p.144 Wind and wave power, Porthcawl, South Wales

Mark Boardman p.27 Hail curtain, Macclesfield, England www.stormhour.com

Kieran Brimson pp.146–147 The frozen beach, Cornwall, England www.kbrimsonphotography.com @kbrimsonphotography

Bill Brooks pp.132–133 The wrath of Storm Eleanor, Newhaven, England www.billnbrooks.co.uk @billnbrooks

Chris Brown pp.134–135 Sunshine and shadows, Kent, England www.my-perspective.co.uk

Marc Brown p.26 Steering clear of trouble, Montpellier, France @pilotseyeview

Stephan Brzozowski pp.166–167 Eden Valley, Cumbria, England www.stephanbrzphoto.com

Amanda Burgess p.156 Cathedral in the marshes, Suffolk, England

Carlos Castillejo pp.148–149 Full house of rainbows, Lofoten Islands, Norway

Debarchan Chatterjee pp.84–85 Kolkata, India www.debarchanchatterjee.com @debarchanchatterjee

Hadi Dehghanpour p.32 Whirling devil, Noush Abad, Iran Hadidehghanpour.com @dehghanpourpix

Dusty Dhillon pp.22–23 Supercell twister, Wray, Colorado, USA www.dustylens.co.uk @dustylensuk

Julian Elliott pp.38–39 Evening strike, Plateau de Valensole, Provence, France www.julianelliottphotography.com @julian_elliott_photography

Michael Epel p.178 A golden layer of fog, Val d'Orcia, Tuscany, Italy

Fereshteh Eslahi p.68 Battling the winds, Bandar Mahshahr, Iran @fereshteh.eslahi

Jaro Fagan pp.154–155 Tsunami of fog, Dingle Peninsula, Ireland @kerryviews

Peng-Gang Fang p.163 Typhoon makes landfall, Kaohsiung City https://www.facebook.com/photorange

John Finney p.44 Devils Tower supercell, Wyoming, USA; pp.58–59 The Enid supercell, Oklahoma, USA www.johnfinneyphotography.com

Andy Fowlie pp.112–113 Arctic smoke, Helsinki, Finland www.andyfowlie.com www.facebook.com/andyfowliephotography

Richard Fox pp.130–131 Stacks of gold, Isle of Lewis, Scotland; p.152 Spectres in the fog, Perthshire, Scotland; p.179 Above the inversion, Devon, England www.richardfoxphotography.com @richard_fox_photography

Julie Fryer pp.104–105 Frost and ice, Cumbria, England

Dianne Kaye Galbraith p.102 Frost in close up, Kangarilla, Australia @galbraithdiannek www.facebook.com/dianne.k.galbraith

Sabrina Garofoli pp.182–183 A river of clouds, Lombardy, Italy

Steven Genesin p.20 Three-pronged strike, Adelaide, Australia @sgeno44

Alan Haynes pp.174–175 Celestial rays, Búðir, Iceland www.alanhaynes.com @alanhaynesphoto

David Hendry p.171 Fog in the valley, Glasgow, Scotland www.500px.com/p/davidshendry @northern_licht

Arron Hiscox p.60 Storm in a spin, Nebraska, USA; p.61 Supercell striations, Nebraska, USA www.seenaturesfury.com @arronhiscox

Patrick Hochner, p.76 Elegance in the rain, Kyoto, Japan https://mykyotophoto.com @mykyotophoto

Vu Trung Huan pp.158–159 Mist in the morning, Phu Tho Province, Vietnam

Jason Hudson pp.114–115 The snow line, Cumbria, England @jasonhudson142

Owen Humphreys p.162 Blue clouds before dawn, Whitley Bay, England @owen.humphreys

Gareth Mon Jones pp.140–141 A sea of clouds, Snowdonia National Park, Wales www.garethmon.com

Boris Jordan pp.24–25 Evening mammatus, Leipzig, Germany. pp.62–63 Lightning on the dancefloor, Frankfurt, Germany @borisjordanphoto

Robert Juvet p.21 Mountain thunderstorm, Appenzellerland, Switzerland https://gurushots.com/robert.juvet/photos www.facebook.com/photoRobertJuvet

Rafal Kaniszewski pp.164–165 A kaleidoscope of colours, Rosendal, Norway www.rafalkaniszewski.pl @rafalkaniphoto

Mikhail Kapychka pp.184–185 Lunar halo, Mogilev, Belarus @mihail_kopychko

Aleksandra Karptsova, pp.180–181 Tidal waves of fog, Vladivostok, Russia @alekskarp

Chirag Khatri pp.172–173 Floating in the fog, Dubai, United Arab Emirates @cvkhatri

Phil Koch pp.92–93 Teeth of ice, Lake Michigan, Wisconsin, USA

Dmitriy Kochergin p.49 After the fire, Bashkortostan, Russia https://kochergin.photographer.ru

Marek Kosiba pp.94–95 The smoking mountains, southwest Bolivia www.marfot.com

Maja Kraljikpp.42–43 Storm on a shelf, Umag, Croatia @majakraljikwx

Ales Krivec pp.168–169 Ring of frost Ratece, Slovenia @dreamypixels www.facebook.com/aleskphotography

Michal Krzysztofowicz p.153 Diamond in the sky, Halley Research Station, Antarctica www.beautifulocean.org

Serhii Kurlia pp.90–91 Aurora at night, Kirkjufell, Iceland @mgbrain www.facebook.com/skourlya

Alan Leightley pp.118–119 A smattering of snow, Northumberland, England www.ilovefridaysme.com @ilovefridaysme

James Loveridge p.170 A waterfall of fog, Dorset, England www.jamesloveridgephotography.co.uk @jamesloveridgephotography

Allan Macdougall p.123 Sculptures of ice, Ceredigion, Wales www.macdougallan.zenfolio.com

Tohid Mahdizadeh p.33 Beating in vain, northwest Iran

Marc Marco Ripoll p.56 Strike at sea, Mallorca, Spain www.marcmarcophotography.com @marcmarcophotography

Mark Mascilli pp.52–53 Anvil of thunder, Oklahoma, USA @mark_mascilli

Jacquie Matechuk p.69 Burning haze, Alberta, Canada www.jacquiematechuk.com @jacquiematechuk

Andrew McCaren p.77 Dam wet, Cumbria, England https://amimages.co.uk/

Shaun Mills p.107 Beach of snow, Mersea Island, Essex, England www.shaunmillsphotography.com

Anne Elizabeth Mitchell p.79 Sunset rolling in, Pisa, Italy www.lizmitchellphotography.com @lizm1tch

Mohammad Hossein Moheimani p.78 Road to nowhere, Agqqla, Iran @mohammad.moheimany

Abdul Momin p.48 Thirsty earth, Chittagong, Bangladesh www.abmomin.com @abdulmomin.bd

Debarshi Mukherjee pp.74–75 A city afloat, Kolkata, India @debarshimukherjee.photography www.facebook.com/debarshi.mukherjee.58

Enric Navarrete Bachs p.57 Lightning by moonlight, Sant Pol de Mar, Spain

Francisco Javier Negroni Rodriguez pp.116–117 Capping the Andes, Patagonia, Argentina @francisconegroni_photographer

Guy Nesher pp.80–81 Flash floods in the desert, southwest Bolivia www.500px.com/p/guyn

Charley Nicholls p.87 The smell of rain, Essex, England www.facebook.com/CharleyNichollsPhotography

Rogier Nieuwendijk pp.110–111 Feathers of ice, Venen, Belgium @rogiernieuwendijk

Tori Jane Ostberg pp.50–51 Red terror, Colorado, USA www.facebook.com/copperstatestormchasing @copperstatestormchasing

Dennis Oswald pp.18–19 Desert planet, California, USA; pp.40–41 Mammatus highway, Texas, USA; p.45 Classic strike, Oklahoma, USA www.dennisoswald.de @dennisoswald

Neil Partridge p.139 Seeing double, Isle of Lewis, Scotland @i.takefotos

Sandro Puncet pp.16–17 Rolling shelf before the storm, pp.34–35 Waterspout party; Mali Losinj, Croatia www.facebook.com/stormchaser.weather.photography @puncets

Dani Agus Purnomo pp.160–161 Supernumeries, Tea fields

Joann Randles p.86 Framed by a rainbow, Swansea, Wales www.JoannRandles.com @joannrandlesphotography

Thomas Rutherford p.145 Wave pile-up, County Durham, England

Elena Salvai pp.46–47 Lighting up the sea, Riomaggiore, Italy

Daniel Sambraus p.97 Shades of blue, Diamond Beach, Iceland www.sambraus.com @danielsambraus

Christoph Schaarschmidt pp.36–37 Strikes on Uluru, Uluru, Australia; pp.100–101, Saxony, Germany www.Christoph-Schaarschmidt.com @hasmonaut

Ruediger Schulz p.83 No place to hide, Saxony, Germany www.photocrowd.com/foxbird

Mikhail Shcheglov p.138 Beauty before the storm, Vik, Iceland www.Photosad.ru

Kolesnik Stephanie p.103 Suspended in ice, Shakhty, Rostov Oblast, Russia www.gurushots.com/Full_Metal_Alchemist/photos

Preston Stoll pp.120–121 Trees of white, Medicine Bow-Routt National Forest, Colorado, USA http://preston.photos

Yuriy Stolypin p.136 Solitary steel star, St Petersburg, Russia; p.157 Piercing through the clouds, St Petersburg, Russia @yuriystolypin

Rudolf Sulgan p.106 Battling through the blizzard, New York, USA www.rudolfsulgan.com @rudolf_sulgan

Bingyin Sun pp.108–109 Clouds stacked up, Vatnajökull National Park, Iceland

Artur Szczeszek pp.150–151 Bridge to nowhere, Bristol, England

Alan Tough p.187 Clouds in the stratosphere, Alloa, Scotland www.flickr.com/photos/7776810@N07

Alexey Trofimov p.96 Sparkling jewels of ice, Lake Baikal, Russia www.en.trofimov-photo.com @trofimovphoto

David Vicary pp.54–55 Lightning shelf, Liberal, Kansas, USA www.seenaturesfury.com @david_vicary_photography

Artur Vinokurov pp.72–73 Coated in ice, Yakutia, Russia @vinokurov_artur

Graeme Whipps pp.176–177 Colours of the aurora, Aberdeenshire, Scotland www.facebook.com/graeme.whipps.photography www.flickr.com/photos/67966149@N04

Lisa Wood p.186 Sunlight reflections Nairn, Scotland www.flickr.com/people/lisa-37-lethen

Tina Wright pp.12–13 Wall of dust Arizona, USA; pp.30–31 A pink dusting, Arizona, USA; pp.64–65 Windmill under attack, Colorado, USA; pp.98–99 Cactuses in the snow, Phoenix, Arizona www.naturesalbum.com @naturesalbum

Yuanyi p.82 Marching on ice, Pinglu County, Shanxi Province, China

Frolova Yulia pp.142–143 A cap of clouds, Vilyuchinsk volcano, Russia